# Reverse Engineering

Edited by
Prof. Dr. Josef Hoschek
Technische Hochschule Darmstadt
and Prof. Dr. Werner Dankwort
Universität Kaiserslautern

B. G. Teubner Stuttgart 1996

Die Deutsche Bibliothek – CIP-Einheitsaufnahme

**Reverse engineering** / ed. by Josef Hoschek and Werner Dankwort. – Stuttgart : Teubner, 1996

ISBN-13: 978-3-519-02633-4 e-ISBN-13: 978-3-322-84819-2
DOI: 10.1007/978-3-322-84819-2

NE: Hoschek, Josef [Hrsg.]



Gesamtherstellung: Präzis-Druck GmbH, Karlsruhe

# Foreword

Looking at modern industrial products, one can recognize a variety of different complex shapes. All these products are not only designed, they are styled. Everybody knows about the importance of styling, if the product is a car, but today even "simple" consumer appliances do not only have to fulfil their function, they must also look nice. In addition, even purely technical products like turbines or valves are designed with very complex shapes to make them work more efficiently.

Thus, optimising the shape of products is one of the key factors in the process chain of development. Today, there are various CAx–systems, which have evolved to be the basic tools for design, calculation, simulation and manufacturing in almost all kinds of industrial environments, but the improvement of the product's shape is - in most cases - done manually on the physical model. This break in the CAD information flow can be overcome with *REVERSE ENGINEERING* techniques reconstructing the shape-describing CAD surfaces (Bézier-, NURBS-surfaces or others) from the modified physical model.

Therefore the 2nd Workshop on current CAx–problems was dedicated to *REVERSE ENGINEERING*. During the workshop were presented

- the newest research results of surface reconstruction for a given set of points
- the methods and tools for problems in Reverse Engineering of some of the most important CAD vendors (Holometric Technology, IBM/Dassault, ICEM, Imageware, Matra Datavision, Tebis).

Additionally, structural aspects in Reverse Engineering, possible future developments and new research directions were discussed.

The participants in the workshop came from France, Germany, Great Britain, Greece, Hungary, Israel, Italy, the Netherlands, Norway, Spain, Sweden and USA. The mixture of researchers, CAD vendors and many technical staff members from automotive companies and their suppliers proved to be very fruitful for a lively and intensive discussion among the the users after each lecture and presentation. The positive working atmosphere in the Bildungshaus St. Michael in Tauberbischofsheim contributed to the success of the workshop.

This volume contains summaries of the lectures, reports on the progress of development of the CAx–systems in the area of Reverse Engineering and a summary of the panel discussion.

In order to compare the Reverse Engineering capabilities of the different CAD systems we have developed some benchmarks (sets of points), which are described in the appendix of the presented proceedings and which are also available on our ftp–server.

W. Dankwort, J. Hoschek
Kaiserslautern, Darmstadt in May 1996

# Contents

# Part I: Presentation of CAD–Systems

# Part to Art

*Sarvajit Sinha, Pradeep Seneviratne*

Imageware

**Abstract**

We posit that free-form design of 3D shapes is far easier when done with a physical medium such as clay than on a 2D computer screen. To this end, we have built a Computer Aided Reverse Engineering system called *Surfacer* which allows conversion of three-dimensional measurements of the physical prototype into NURBS surfaces that can be manipulated by most Computer Aided Design and Manufacturing systems. The system's interface has been made simple enough to be mastered by anyone, and the software provides a number of mathematical algorithms that simplify the task of converting 3D data into surfaces. This software and its algorithms are the focus of this paper.

## Introduction

The shape of finished goods and manufactured products today exhibit a marked emphasis on smooth shapes and free flowing contours. Whether it is cars, audio equipment, cameras, ergonomic furniture or injection molded plastic products - all have smooth contours in their design. However, the production of manufactured objects from such shapes is not an easy process. To this end, a geometric modeling software can provide the control and flexibility to create free form contours in the design and manufacture of products. However, it is still a laborious process.

The lack of effective software and processes has created an unwieldy process of 3D design on a 2D medium (a computer screen) and slowed the integration of design and manufacturing environments. The link between the physical object and its CAD representation lies in 3D scanners and user-friendly software environments in which one can transform the data produced by these sensors into a CAD acceptable and, ultimately, manufacturable form.

One of the major bottlenecks in current CAD systems is the inefficient representations and manipulation tools for the design of free-form, sculptured, three-dimensional shapes. We propose a bold and perhaps devolutionary approach that is at once backward and up-to-date in its approach. The approach is based on a presumption that the definition of irregular, sculptured 3D shape is difficult, if neigh impossible, to capture in the 2D environment of a computer screen. We, therefore, propose that artists and skilled tradesmen be allowed to design objects the way they always have- i.e. by building prototypes in clay, foam or wood. This frees up the skills of the artists to concentrate on tasks in the creative process at which they excel. The task of converting this 3D prototype shape into a computer model for subsequent CAD/CAM/CAE processes, *or part to art*, is then left to the computer and specialized software.

In this paper, we show how the creation of physical prototypes, and recovery of geometry from this prototype in a semi-automatic process, has be coded into a process that is simple and efficient with the *Surfacer* system.

### Reverse Engineering and Part to Art

Reverse engineering in its complete definition refers to the process of create a completely engineered prototype, a clone in form and function, from a physical part. In this paper, we will utilize a narrower definition of the term to refer to the process of recreating the form or mathematical CAD model of the part from a set of measurements of it. Such a process is already used in industrial design studios, where designers first work in a physical medium such as clay or foam before creating a CAD model of a part. This is the case in the automotive industry, and also true in the aerospace, shipping and ancillary industries and in the consumer product design. Reverse engineering is also being applied to the manufacture of one-of-a-kind prosthetic devices, where the artificial limb must fit the amputated limb snugly.

Reverse engineering depends upon a sensor to collect measurements of the part, and software to convert the data into a CAD model. The next two sections delve into these areas.

## 3D Sensors

3D sensors are broadly classified into two types, contact sensors which touch the object in order to take measurements of it, and non-contact sensors. Contact sensors such as Coordinate Measuring Machines are powerful and sophisticated and are already common in industry as devices for measurement or inspection of parts. These consist of an arm or gantry mechanism with a touch-probe mounted on an effector. These contact sensors collect data by contacting the part at every measurement location. Hence, data collection is a slow and laborious process. Due to the mechanical nature of these devices, they are also cumbersome to use. For highly accurate devices, they can also be very expensive.

Non-contact Laser or light based systems (which includes Moiré technology) provide an easy and fast method of acquiring 3D information about objects. Overviews of the underlying technology behind sensors can be found in Besl (1988) or Wohlers (1995). These sophisticated measurement systems furnish accurate dense point measurements very quickly. The availability of this point data allows new powerful design and manufacturing processes, which reduce time and improve quality. With improvements in these sensors, it is possible to employ them in a number of areas where they were previously considered esoteric. For example, engineering design, inspection and verification are some of the manufacturing applications to which they can be employed.

## RE Software

Unfortunately, there is a significant gap between the data that these sophisticated sensors provide and geometric information that can be used by subsequent design and manufacturing processes. Researchers such as Sarkar and Menq (1989), Sinha and Jain (1993), have explored automatic techniques to convert raw 3D data into a form which is acceptable to CAD/CAM systems. Since completely automatic general purpose techniques are unlikely to be feasible in the next decade, we believe that two technologies are essential for the integration of reverse engineering software in design and manufacturing environments. They are:

I. A flexible, interactive software environment- An appropriate software environment is needed in order to facilitate handling and processing of large amounts of data.

II. Surface Creation by fitting: Mathematical techniques such as surface fitting for converting the data acquired using the sensors to appropriate geometric entities.

In this paper, we will show how *Surfacer* solves these problems. With respect to the software environment, conventional Computer Aided Design (CAD) systems present two major problems. One problem is that they lack the capability to handle the large point data sets created by 3D scanners. Typically, CAD systems treat a point as an XYZ triple, which has the same importance as other geometric entities such as lines, arcs or surfaces. Thus each point is an "object" within the system, with auxiliary information such as name and color stored in it. The storage of such redundant information places an unnecessarily heavy burden on computer memory resources, particularly when the number of points climbs into the 100s of thousands. Within *Surfacer*, we treat the entire collection of points as a single object which is called a "cloud".

The second problem with CAD systems is in their inability to provide a complete reverse engineering environment. This problem is solved by Imageware's *Surfacer*. It offers a user-friendly software environment for visualization, and manipulation of point data and surfaces for design and manufacturing applications.

In the following sections, we will describe the unique features I and II which *Surfacer* affords.

### I. The Software Environment

#### a) Data and geometry visualization

Through visualization modes and display options *Surfacer* gives the user myriad means of viewing any a piece of geometry. Points collected from a sensor can be displayed as pixels, crosses or polylines. *Surfacer*'s triangulation algorithms allow the user to create a polygonization over the data and visualize the data in a shaded manner. Dense point sets form a complete representation of the object which we refer to as the "virtual model". Figure 1 illustrates the difference between the visualization of point data as individual range elements and as shaded polygons (the virtual model). The shaded model conveys a great deal of information about the shape of the object, a fact that is heavily relied upon while reverse engineering. This is the essence of the virtual model, since it brings out the flow of curvature over the sculptured surface, a key capability in trying to establish patch boundaries. The advantages of *Surfacer*'s polygonization algorithms are that they operate on even unorganized point sets. Figure 1. above illustrates this visualization capability.

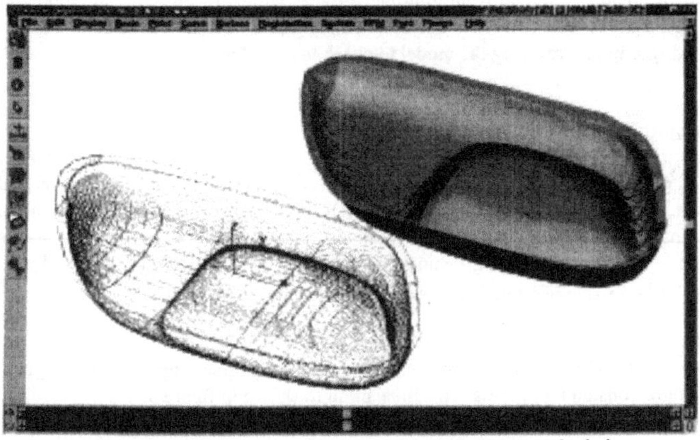

Figure 1. Visualization of a point cloud as scattered points and as a shaded polygonized "virtual" model.

b) **Data Set Editing**

*Surfacer* provides about 60 different tools editing point clouds. For example, the user can edit points by

- Selecting a subset of a cloud, thereby manually removing data points that are gross outliers
- Selecting a subset of a cloud to work on.
- Cutting cross-sections through the cloud.
- Smoothing the data using traditional filters

These tools can eliminate both Gaussian noise and outliers, reorganize the data into section lines, automatically extract feature lines, and partition the data into appropriate faces of the object. Figure 2 illustrates two simple point selection capabilities- that of roping off a set of points to create a sub-cloud, and of "cutting" cross-sections through a dense cloud to create axes-oriented section curves.

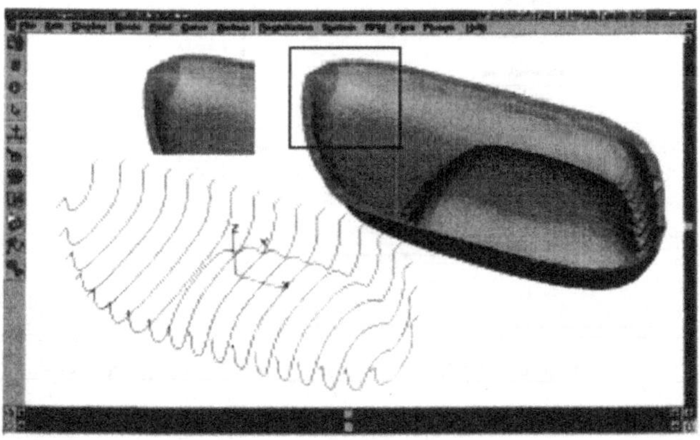

Figure 2. Point subset creation functionality

**I. Surface creation**

This section touches upon the complex process of creating *multi*-surface models from point cloud data. While most academic papers describe algorithms to create single surfaces or patches, the real challenge in an industrial setting is one of creating a model from multiple surfaces while ensuring that the model as a whole is completely smooth. In the sections below, we will describe the algorithms needed for surface creation, ensuring continuity of the model, and finally interrogating the model to check these values.

**Representation**

Whereas the problem of surface representation has been largely resolved for "regular shapes" ), the representation of free-form sculptured shapes is still an open question. While the former, such as those modeled by a solids-based modeling systems, use a mixture of NURBS and implicit equations, the latter has basic forms of Gordon, Coons, Bezier, and NURBS surfaces all vying for acceptance. We use only NURBS for our representation. The main reasoning behind this choice is that NURBS are universally accepted within the CAD community. Surfaces generated by the system are output in a variety of CAD compatible formats, including IGES and VDA.

**Process**

When traditional CAD systems are applied to reverse engineering from scan data, they operate within a complex and time-consuming process. This the paradigm shown in figure 3.

The process of building, interrogating and refining each curve and surface is much akin to the compile-link/test/debug cycle of creating computer code and is an art more than a science.

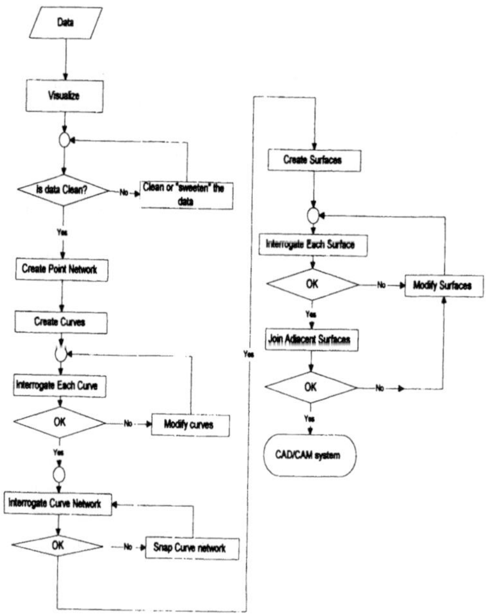


Figure 3. Traditional "surfacing" in reverse engineering

Surface modeling in a commercial CAD system consists of defining a set of patch boundaries or a curve network describing the shape, and then lofting, blending or sweeping these curves to obtain the skin of the object.

Algorithmically, the process of blending 4 bounding curves to create a surface (figure 4) consists of first

- making the *u* and *v* parameter curves "compatible" i.e. making their knot vectors the same by inserting knots at appropriate locations in the *u-pair* of curves and in the *v-pair* of curves and
- then applying a bilinear or bicubic blend between the two curve sets.

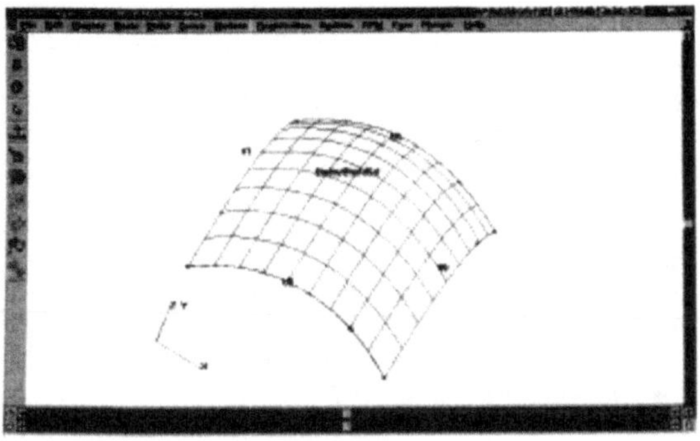

Figure 4. Blending 4 curves to obtain a surface

The equations for creating surfaces from curves can be found in Piegl(1988) or Farin(1988).

While building curve networks, one must ensure that curves are smooth and continuous. Therefore one needs to ensure that the tangent, curvature and sometimes the 3rd derivative of the curve are well behaved. This is another drawback of the current reverse engineering process since the user has to spend a great deal of time fixing entities (curves) that are only incidental to those he really wants (surfaces).

Typical problems in creating surfaces from a curve network are:

1. Very close knots. These are typically caused by the compatibility process mentioned above and lead to numerical instabilities.
2. Irregular iso-parameter flow. This condition is caused by curves not being parametrized by chord length.
3. Discontinuities between surfaces.
4. Small radius of curvature or crimping in the blended surface, which lead to a non-manufacturable surface.

When such problems occur, the user usually has to backtrack all the way to the curves that form the network and fix the problem there. This is a most time-consuming loop in the process.

Another major drawback of the current process of reverse engineering is that while the curve mesh or network is built from the data points, the surfaces built from the curves do not have any correlation to the data. The surface shape is completely defined by the blending function used in the process described above. Therefore, there is no

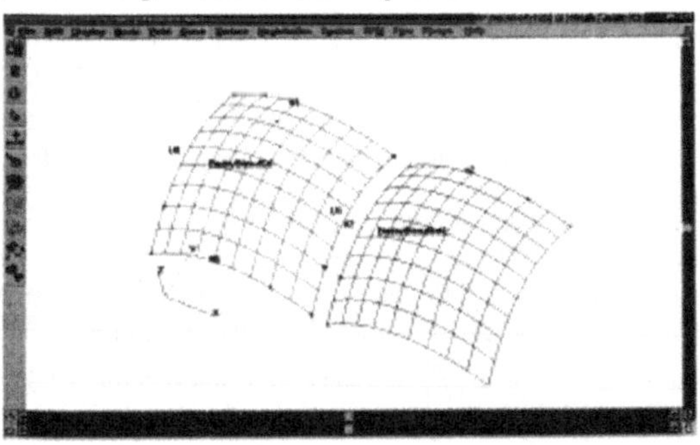

Figure 5 Parametric directions across two adjacent patches

hope of creating surfaces within any required tolerance unless one increases the density of the curve mesh to achieve accuracy. More curves mean the creation of more surfaces, and this obviously implies more time.

While users of *Surfacer* have the capability to fall back on the traditional methods of surface creation described in this section, the *Surfacer* has introduced two important new concepts and dozens of tools which dramatically speed up surfacing. These are described in the next two sections.

**a) Surfaces from data**

The heart of our system lies in a powerful surface fitting engine which accepts data in arbitrary form. The data does not have to be arranged in a grid or measured along character lines. The engine then converts it into a compact, precise mathematical description. Existing surface fitting algorithms include a variety of different approximation schemes. These schemes include, for example:

- simple bivariate polynomial approximation -- planes and low-order polynomials;
- the standard set of quadrics -- spheres, cylinders and cones (in order to recover the basic primitives of solid modeling systems); and
- B-Spline approximation for modeling free form surfaces.

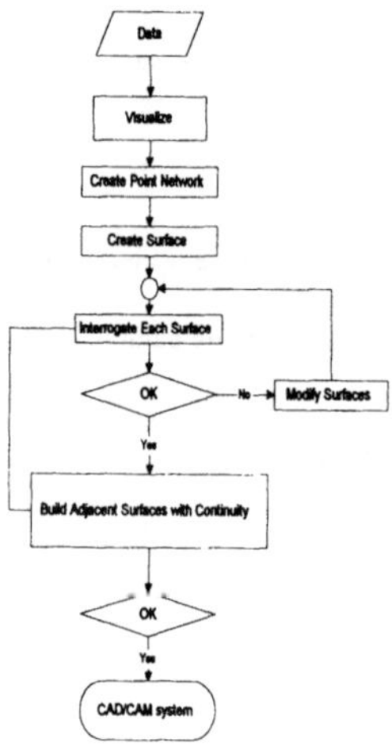


Figure 6. Reverse Engineering process in Surfacer

In this paper, we only mention B-spline fitting methods. There are a number of optimization-based algorithms available (e.g. Schumaker, L. L., 1979, Sarkar, B. and Menq, C. H., 1991, Hoschek, J, and Lasser, D., 1993, P. Lancaster and K. Salkauskas, 1986) that allow the user to "fit" in some norm-minimizing sense, a B-Spline to the data points. These approximation schemes find a surface that minimizes a functional $\Im(f(u,v))$ that is a weighted sum of the square of the error $E(f)$ and some smoothness norm $S(f)$. The error is computed between the data and the fitted surface $f(u,v)$

$$\Im(f) = E(f) + \lambda(S(f)$$

where

$$E(f) = \sum_{i=1}^{N} (f - p_i)^2$$

and

$$S(f) = \lambda_1 \iint_{\Omega} \left( f_u^2 + f_v^2 \right) dudv + \lambda_2 \iint_{\Omega} \left( f_{uu}^2 + 2 f_{uv}^2 + f_{vv}^2 \right) dudv$$

Here $f = f(u,v)$ is the approximating function, $p_i, i = 1..N$, are the cloud points, and $\lambda_1, \lambda_2$ are weighting coefficients. In this formulation, the surface $f(u,v)$ is represented by a B-Spline

$$f(u,v) = \sum_{j=1}^{J} \sum_{k=1}^{K} c_{jk} B_j(u) B_k(v)$$

where *B()* are the standard B-Spline basis functions, $c_{jk}, (j = 1..J, k = 1..K)$ are the control points of the model, and $u,v$ define the parametrization of the surface. More details can be found in Seneviratne and Sinha (1993) or Sinha (1991).

The smoothness norm $S(f)$ shown above is referred to as the thin-plate norm, and is just one example of norms that can be used for fitting. It is essentially a measure of the consistency or bending energy of the surface and is related to the variation in the normal of the surface. The thin-plate norm is an approximation to the curvature of the surface for small derivatives, and its primary usefulness lies in the fact that it's solution leads to linear systems of equation. A smoothness functional based on the curvature or minimum variation, such as that by Moreton (1992), leads to non-linear systems of equations with the adherent problems in solution.

Another important property of this procedure is that it eliminates high-frequency or rapid variations in the data, and produces a smoothly varying surface. Surface fitting can thus be viewed as a low-pass filtering procedure, and removes the need for data "sweetening" or smoothing as the process of smoothing erroneous data collection.

An important question to be answered is the creation of the parametrization of the surface, i.e. the choice of the knot vectors $u_i = \{u_0, u_1, \ldots. u_m\}, v_i = \{v_0, v_1, \ldots. v_n\}$ in the parametric directions. The parametrization of the surface can be rectangular, based on boundary curves, or dependent on another surface. If the data is single-valued in some direction, the surface function is represented as $f(x,y)$ and the points $p_i$ represented by the value $(x, y, z_i)$. On the other hand, the data set may not be single-valued, in which case the data points are mapped onto some pre-existing parametrization and the function is $f(u,v)$ and the data points are $(u, v, w_i)$. Seneviratne and Sinha (1993) elaborate further on this point. The surface fitting functions have been developed for Cartesian as well as cylindrical coordinate systems.

The quality of the fitted surface $f(u,v)$, that is how closely the surface resembles the point cloud depends on the parameters $\lambda_1, \lambda_2$ chosen. For a vast majority of data sets one can obtain a quality surface fit by wisely selecting the values of the surface fitting parameters.

**b) Building with continuity**

In creating a water-tight, continuous surface model needed for manufacturing, the creation of the CAD model of a part requires the creation of a multi-surface skin. The models must be smooth *within* patch boundaries, as well as *across* adjacent surfaces. Multi-patch models should be seamless in position and in derivative values across seams. The seam should be completely invisible when interrogated with a *reflectance* or the *curvature* plots of adjacent surfaces. Every curve and surface patch must be created carefully, interrogated, tweaked if needed, and re-interrogated, until a perfect entity has been obtained. The user has to pay a great deal of attention to boundaries, since this is where many problems arise. Once the user has completed a single surface to satisfaction, he or she can concentrate on the neighbors. Earlier versions of *Surfacer* offered tools to stitch two surfaces together, and thereby ensure the continuity conditions between surfaces. The current version of *Surfacer* allows the user to build surface models with inherent continuity.

In order to achieve visual or $G^1$ continuity, the necessary boundary conditions between patches are:

$$\frac{\partial(f_1)}{\partial(v_1)} + \alpha(u_1)\frac{\partial(f_1)}{\partial(u_1)} = \beta(v_2)\frac{\partial(f_2)}{\partial(u_2)}$$

where

$f_1(u_1, v_1), f_2(u_2, v_2)$ are adjacent surfaces and $u_1 = v_2$

Note that when $\alpha = 0$, $C^1$ continuity is achieved.

In order to "build with continuity", the surface fitting equations of section IIa have to be modified into a constrained optimization formulation. The constraints of position/tangent/tangent-plane continuity (above) are built into the fitting equations directly.

Building with continuity is not the same as a build followed by stitching. If the user were to build geometry with no boundary constraints, he would get inadequate surface edges. This is very easy to do because no boundary constraints means the boundary region is free floating. Stitching will only magnify the problem by affecting neighboring surfaces. If the user were to build with continuity, free floating edges would not allowed (except for the seed surface), and so the results would be much better.

Build-with-continuity avoids the problem we call "chasing-your-tail-around". If you have three or more pre-built surfaces meeting at a vertex, fixing continuity by stitching the first two would have meant that continuity with the third would be lost. In the new scheme, continuity is built in at the time of creation, and surfaces already created are not altered because of a master-slave constraint.

**III) Analysis**

In order to provide a complete solution to users, *Surfacer* not only provides the facilities for model creation as described above, but also for interrogating or analyzing the models to ensure that the models are smooth and continuous. The tools available in the Geometric Analysis sub-system of *Surfacer* allow the user to easily complete a host of verification tasks. These tools include (but are not limited to) finding distances between a point and the surface, finding the difference between a cloud of points and a surface; the statistical analysis of the errors associated with the fit (mean, standard deviation etc..). Once the curves or surface have been created, one can check the degree of continuity between adjacent surfaces as well as interrogate adjacent surfaces using tangent, reflectance, normal or curvature plots. See figures 7 and 8 for details. These are typically main-stream CAD functionalities, but are offered in *Surfacer* so that the user can complete his complete project within the environment provided. An overview of interrogation tools for surfaces can be found in G. Farin(1990).

If these tools reveal that the fit was poor in either accuracy or smoothness, the user can adjust the fitting parameters to obtain a better fit. If repeated attempts of changing the fitting parameters do not produce a suitable fit, one can use advanced surface editing techniques available in the *Surfacer* environment. These are not described in this paper.

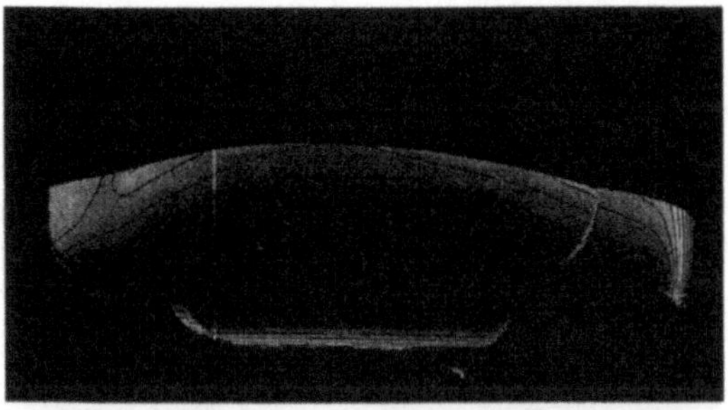

Figure 7. Free form surface from automobile exterior with parts covered with a reflectance map.

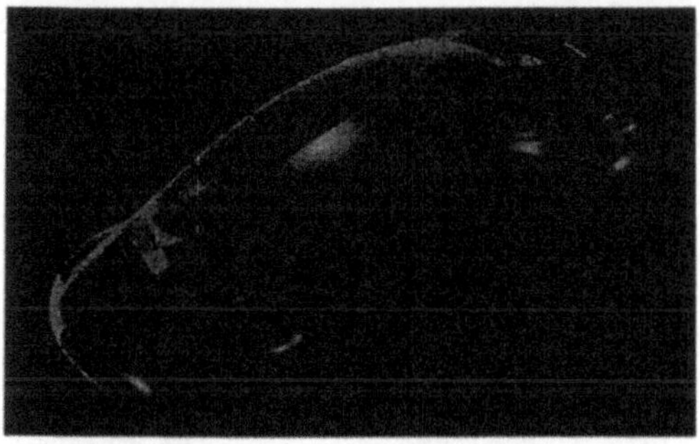

Figure 8. One half shows the original data, a range image of a free form surface from the body panel of an automobile. The second half shows the output of a spline creation procedure.

As another example, we present figure 9. This shows the original data (a), the surfaces created from the data(b), (c) and a representation of the accuracy of the fit(d). This model was created in matter of minutes from the original data.

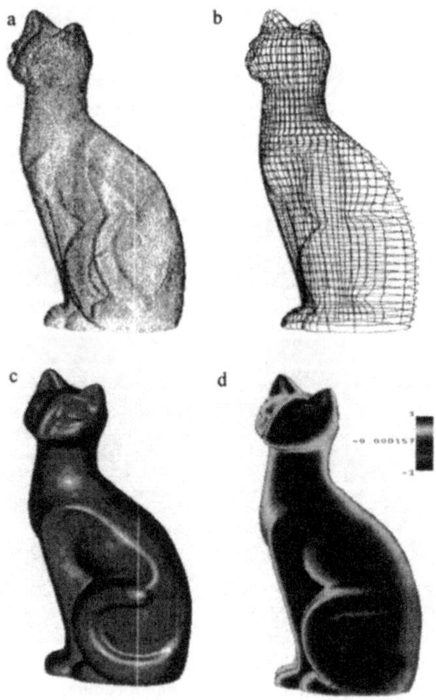


Figure 1. a. Original Data b. Surfaces displayed as a mesh c. Shaded surfaces d. Difference Analysis

Finally, we present some of the benchmarks that were presented at the Tauberbischofsheim conference on Reverse Engineering.

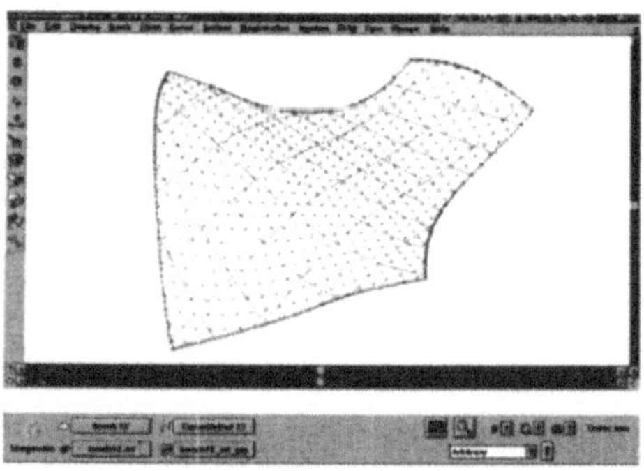

Fig. 10. Benchmark 1. Surface fit to data and trimmed back.

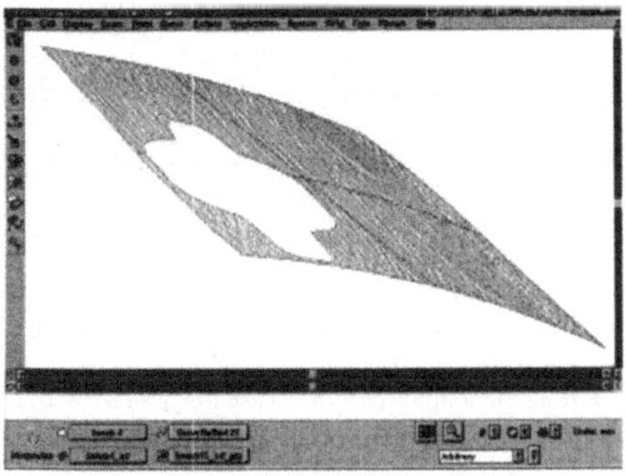

Fig 11. Surface fit and trimmed to data.

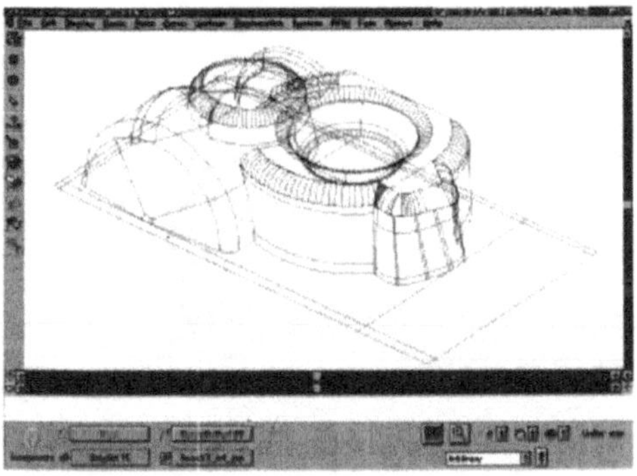

Fig. 12. Multiple prismatic surfaces fit to data along with fillets created at borders

## Conclusions

Reverse engineering stands to turn the problem of the design of sculptured free-form surfaces completely around. We espose a part-to-art paradigm which reverts back to some old techniques(sculpting) and some new techniques (constrained non-linear optimization). This paper claims the creation of free-from shape models using the techniques described in this paper. While the new process still requires a certain amount of interactivity, user interaction is greatly reduced by the application of combinatorial and optimization based methods.

The process of reverse engineering has been taken out of the research laboratory and made into a viable commercial product. It has been validated by hundreds of users of the product who are actually getting prototypes to market faster by this process. Some sample results of the reverse engineering paradigm for designing sculptured surfaces have been presented in this paper.

## References

Besl, P. J. 1988, "Active optical range image sensors," Mach. Vision and Applications, Vol. 1, 127-152

Wohlers, T., 1995, "3D Digitizing for Engineering," Computer Graphics World, March, 1995.

S. Sinha and R. Jain, 1994, Range Image Analysis, In Handbook of Pattern Recognition and Image Processing: Computer Vision, Vol. 2 pp. 185-237, Academic Press.

Seneviratne, P., and Sinha, S., 1993, Single-valuedness, parametrization and approximating 3D surfaces using B-Splines, SPIE Vol. 2031, Geometric Methods in Computer Vision II, pp. 193-204.

Sinha, S., 1991, Surface Reconstruction in Computer Vision, Ph.D. thesis, The Univ. Of Michigan, Ann Arbor.

Schumaker, L. L., 1979, Fitting surfaces to scattered data. In "Approximation theory II" ( G.G. Lorentz, C. K. Chui, and L. L. Schumaker eds.), pp. 203-268. Academic Press.

Sarkar, B. and Menq, C. H., Smooth Surface Approximation and reverse Engineering, Computer Aided Design, Vol. 23, pp. 623-628.

Hoschek, J, and Lasser, D., 1993, Fundamentals of Computer Aided Geometric Design, A.K.Peters Press.

P. Lancaster and K. Salkauskas, 1986, Curve and Surface Fitting, Academic Press.

R. Franke and L.L. Schumaker, 1987, Bibliography of multivariate approximation, In Topics in Multivariate Analysis, pp. 275-335.

G. Farin, 1990, Curves and Surfaces for Computer Aided geometric Design, Academic Press.

Piegl, L. and Tiller, W., Curve and Surface construction using rational B-Splines. Computer Aided Design, Vol. 19 pp. 485-498.

V. Krishnamurthy, 1996, Fitting Smooth Surfaces to Dense Polygon Meshes, pre-print, Stanford University.

Henry P. Moreton, 1992, Minimum Curvature Variation Curves, Networks, and Surfaces for Fair Free-Form Shape Design, Ph.D. thesis, U.Cal., Berkeley.

## Authors

Dr. Sarvajit S. Sinha
Imageware
313 N. First Street,
Ann Arbor, MI 48103
sss@iware.com

Pradeep Seneviratne
Imageware
313 N. First Street,
Ann Arbor, MI 48103
pradeep@iware.com

# Surface Reconstruction

*Heinz Rybak*

Matra Datavision

*As follows the main steps and the methods of surface Reconstruction are described with the example of the CAD system STRIM and its dedicated application modules. With it the advantages of special function in combination with basic CAD functionality is shown.*

## *Objectives*

*The main objectives of Surface Reconstruction is the building of CAD models out of point files. Those files are either*

- *output from mechanical or optical digitization of a mock-up or old pattern*
- *a set of points computed by specialised application software (turbine blades)*
- *digitization of non manufactured objects (human body parts, work of art and so forth)*

*At first a precise description of a model needs the generation of a large set of points. Therefore the processing of points seems to be time consuming and doubts may come up whether Surface Reconstruction is an appropriate method for the creation of CAD models.*

*A closer look shows that several reasons speak for Surface Reconstruction:*

- *It is cheaper : the building of CAD models building from scratch still remain in any case a long task which moreover needs skilled user.*
- *Furthermore it is a quick way to design a component embedded in a complex environment (gas tank)*
- *It is still the best way to evaluate ergonomics like fitting with human body (seat, handle of fly stick, etc.) In these cases mock-ups are still required.*
- *It is easier to use than a classic CAD surface modeler: process does not start from scratch, the set of points is used as a base to model from.*

## *Areas of application*

*From these advantages specific areas result for the application of Surface Reconstruction. In general this method is used for*

- *the capture of patterns or mock-ups in the CAD system*
- *use of models supplied in the form of digitized points*
- *recovery of manual modifications carried out on a prototype model*

*In particular one can find a need for this technology in following branches*

- *Toys industry: children masks, shaped plastic toys, ..*
- *Sport industry: recovering of the CAD model of mock-ups articles like golf clubs, crash helmets, ski boots; mock-ups are required for styling and for ergonomics (fitting with human anatomy)*
- *Medical products: modelling of human environment so as to custom-made-to-measure equipment (prostheses, shoes)*
- *Consumer products: vacuum cleaners, razor, iron, bathroom articles, . . . .*
- *Bottle / packaging industry: mock-ups are required for styling and for ensuring fitting with the hand anatomy*
- *Furthermore some perfume bottles are to complex to model easy from scratch with a CAD system, it is better to start with a mock-up.*
- *Jewellery industry and reproduction of works of art: jewels are often designed with mock-ups or no initial model of a work of art is available.*

## *Surface Reconstruction as part of the Design Process*

*The following schema shows the general procedure using Surface Reconstruction as Method for the creation of models. It has to be regarded that the building up of a CAD model is mainly an iterative process, even though it is not shown so to simplify matters.*

*The modelling of the surfaces is often supported by a machined model (e.g. for the creation of visible shapes of consumer products). In addition the use of Surface Reconstruction in context with the modification of the machined model is shown. This procedure is necessary if modifications of a tool are required due to the manufacturing process for instance. Especially such update of tool modifications is additionally an important area for Surface Reconstruction.*

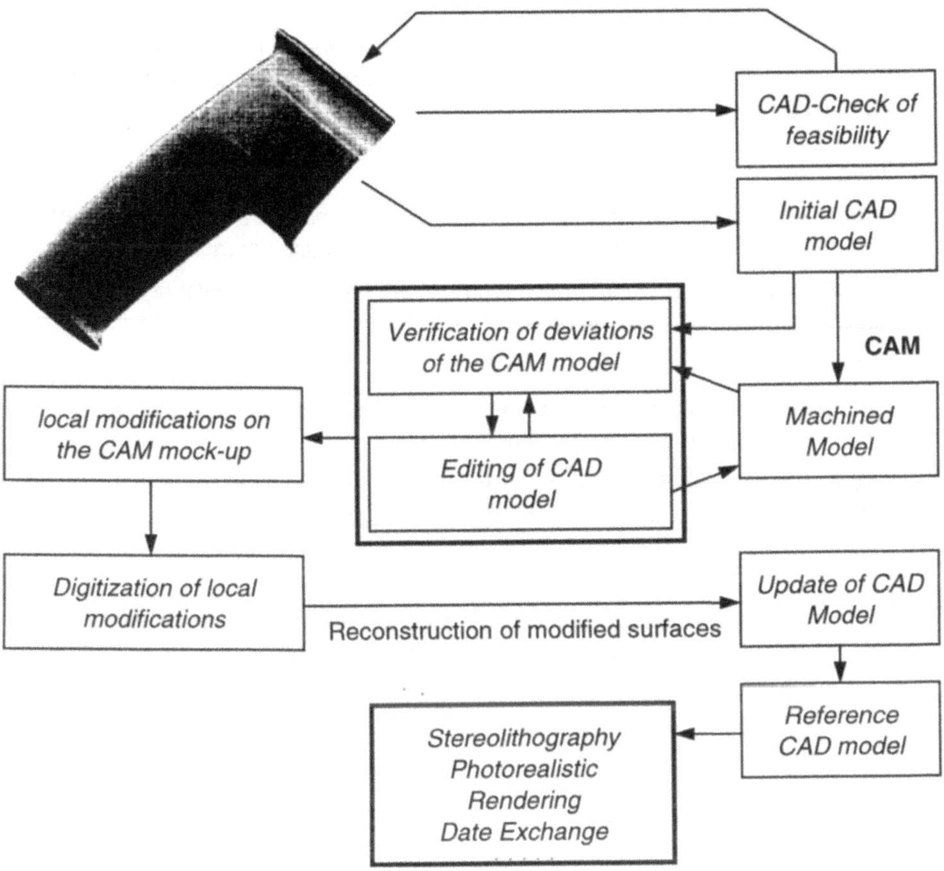


## *Requirements*

*In spite of the advantages described above and the wide and different range of application the stylist and the designer have specific requirements with regard to various characteristics:*

*Productivity:*

- *The point capture itself has to be carried out automatically or batch processed. Subsequently specific methods must be available to process those sets of points.*
- *The CAD system as basis for the surface reconstruction has to enable the user to minimise the number of entities concerning surface degree as well as concerning the amount of surface itself.*

*Independence:*

- *According to the method and means of digitization (mechanical, optical or others) overall scanning has to be possible. This is a prerequisite for a subcontracting of the point capture.*

*Accuracy / quality:*

- *In order to respect the part and the designer ideas character lines, high curvature area and surface boundaries should be easily recognised or modelled.*
- *Smoothing process and interpolation processes have to be available to clean the models flaws and to avoid wrinkles.*
- *Surfaces have to be tangent to each other.*
- *Deviations between the physical model and reconstructed surfaces should be controlled with a set of tuning parameters.*
- *Finally adapted checks are needed to find deviations in relations with digitized points and for quality analysis for the generated surfaces.*

*Flexibility:*

- *All CAD modelling functions of the CAD system below have to be available at any time. In other words a Surface Reconstruction Application has to be embedded in a CAD modelling system.*
- *All curves created with Surface Reconstruction shall to be used directly as surface modelling input.*

## *Main Reconstruction Steps*

*As follows the main steps and the methods are described with the example of the CAD system STRIM and its dedicated application modules. With it the advantages of special function in combination with basic CAD functionality is shown and how the above mentioned requirements can be fulfilled. The example for demonstration is a part of the outer shape of a drilling machine*

- ***Digitized Files Pre-processing***

*Normally point files can not be directly processed. Often several files coming from several digitizing operations have to be merged first. As shown with Picture 1 rendering of the raw date file helps checking defect, forgotten or overlapping areas.*

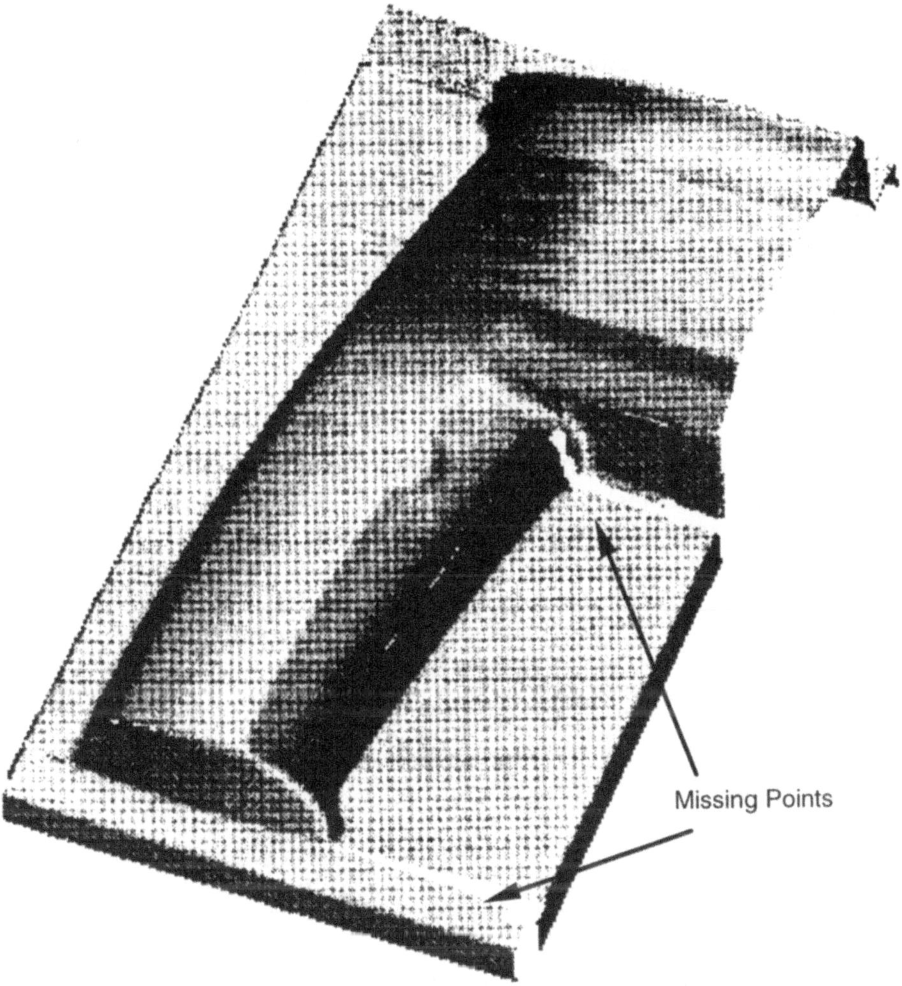


Picture 1: Rendering of measured points

*To edit these points among others following functions are available:*

- *Trimming of certain areas,*
- *Elimination of invalid points,*
- *Offset of digitized areas,*
- *automatic detection of sharp edges or characteristic lines and*
- *filleting of open areas with a given radius.*

*Three of this editing function shall to be explained as follows.*

*As Picture 2 shows any closed contour can projected onto the digitized area. Only the points outside the domain are kept. The parts of lines deleted a replaced by a segment. This function is useful if certain areas of the digitized part are well known or shall fulfil defined conditions.*

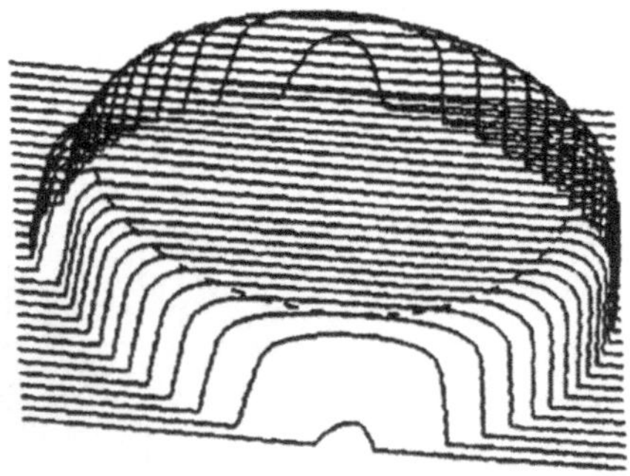

Picture 2: Trimming of points

*A further method for the modification of digitized points is the use of given or user defined characteristic lines. Provided that the line is located in the digitized plane all point within a given distance are deleted and new point are created onto the characteristic line. As it is shown, this approach can be applied to two directions.*

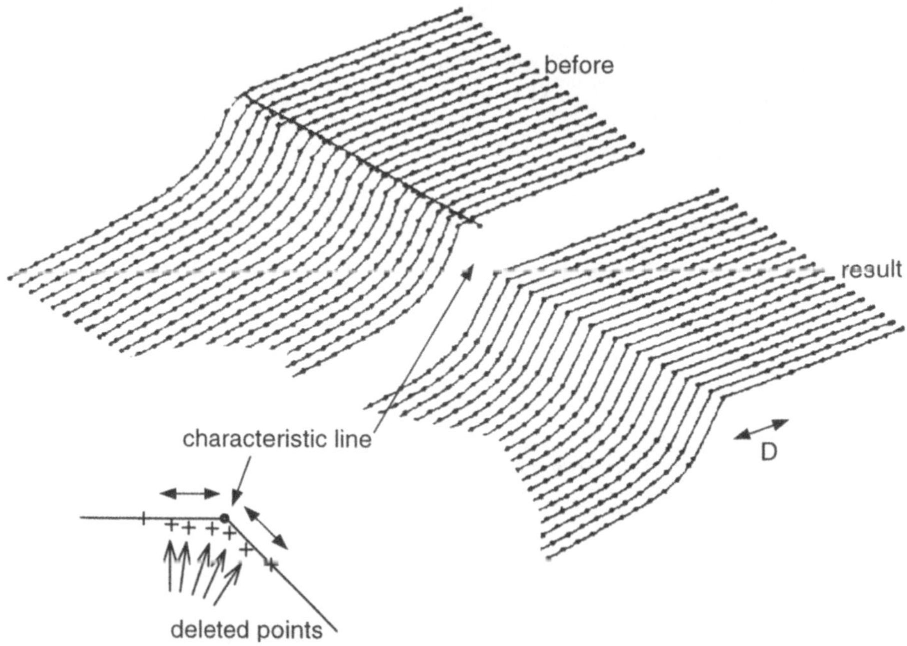


Picture 3: Modification of points by a characteristic line

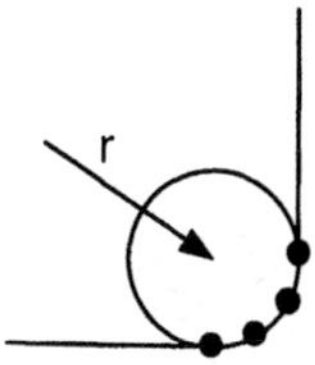


Picture 4:
Filleting

*Another helpful function is the filleting of open areas (Picture 4). This is important if unknown blending areas shall filled with points (see also Picture 1). For that purpose a radius is given and the system creates automatically the missing points. This function can be used either for certain areas or for the whole set of given points if necessary. So missing point can created very easily.*

*The described function are helpful to clean the digitized point. As side effect the number of points may be reduced. This is necessarily not the main goal: Experience shows that data should be as dense as possible. It is easier to rebuild a model from too many points than too few.*

- ***Grids Generation***

*Provided that all points are corrected so called grids are generated (Picture 5)*

*This grids are the basis for the following automatic creation of curves. The smaller the grids are the better the accuracy of the curves created later. On the other hand a high amount of date reduces the computation performance, if the grids are too small. Finally It depends on the experience of the user, which values result in good surfaces.*

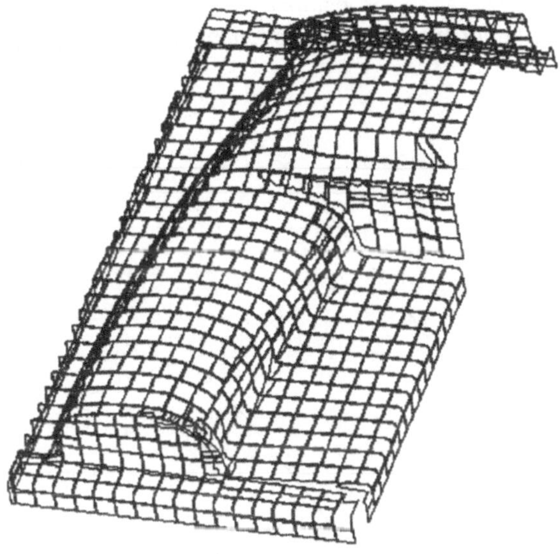

Picture 5: Generation of Grids (presentation of borders)

*The grids are used as described with Picture 6.*

- *At first the projection of a curve (as shown) or the cut by a plane is projected onto each single grid.*
- *Then a group of curve points is created for each grid.*
- *At third the curve points are merged and used for the creation of a single curve, which fits the digitized points as good as possible.*

- *Alternatively the curve points of a grid can be used first for the creation of a single curve. The curves of all concerned grids are merged afterwards for the building of a composed contour.*

*This general method is base of all different approaches for the creations of curves described as follows*

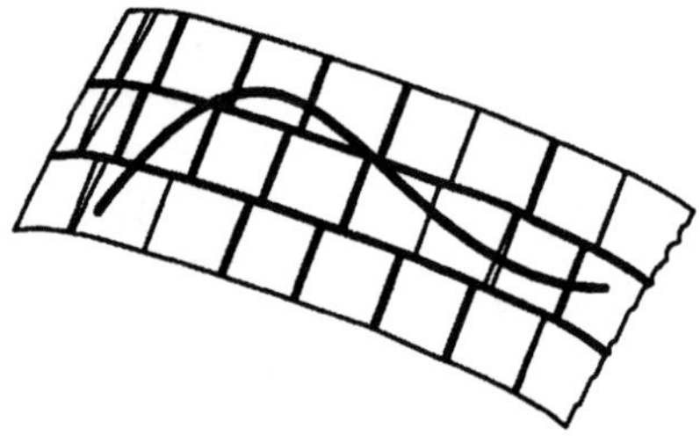

Picture 6: Projection of a curve onto a grid assembly

- ***Character Lines Generation***

*Next step is the generation of Character Lines. A simple digitizing operation by parallel planes for instance is not generally enough to give information about the Characteristic Lines or discontinuity lines of the part. Therefore different methods can be used to define those lines.*

*Following approaches are possible:*

- *automatic approach*
- *controlled approach (3D curves freely created on the set of points and dynamic planar curves designed from the user view point, then projected on the digitized points)*

*The creation of the main section (Picture 7) is an example for the automatic approach: it is a cut between the digitized points and a plane.*

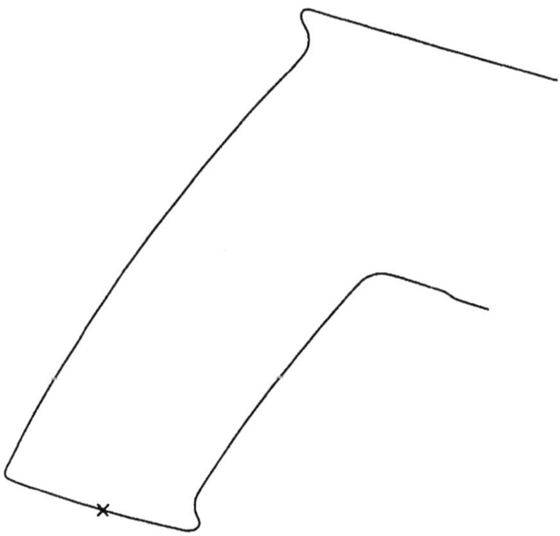

Picture 7: Creation of the main contour

*In a similar way Picture 8 shows at first contours, which are created by simple cutting with planes. At second the cutting planes are defined by a driving contour and user defined points. In this case the planes are perpendicular to the driving*

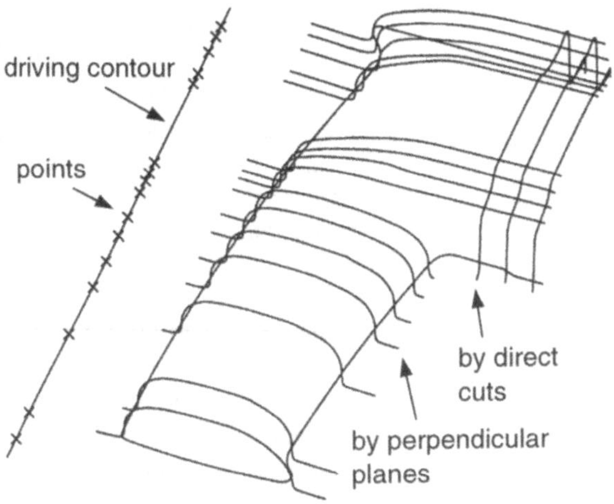


Picture 8: Automatic creation of contours

*contour. In both cases the contours are created on base of the previous defined grids (see Picture 5 and 6). The system tries to define for each grid one single curve, which is as near as possible to the digitized points. The complete contour consists of the sum of all single curves. As mentioned above it is obvious that the accuracy depends on the size of the grids.*

*A further method is the direct creation of a 3D contour by using certain digitized points (Picture 9). After selection the system tries again to build single curves an base of the grids and assembles all curves to a contour.*

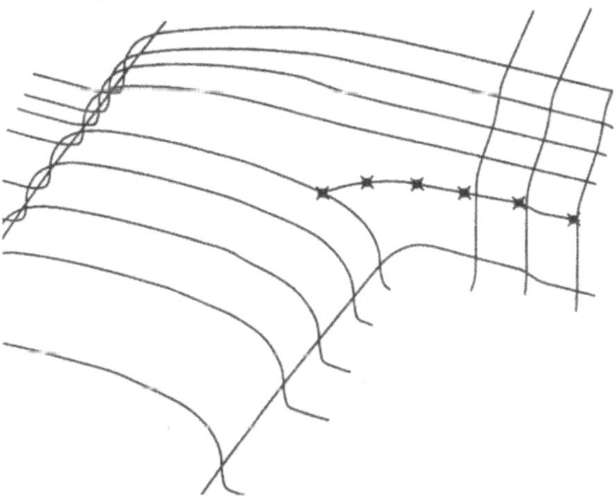

Picture 9: Creation of 3D contour

*Picture 2 shows now, that the used demonstration part has a boss. To obtain a good final surface quality it is useful to define a boundary around. For that first a voctor normal to the previous defined grids is created (Picture 10). A second normal vector is positioned in the steeply sloping area of the boss. With both normals finally the boundary curve is computed. This method is extremely useful. Especially sharp edges can be easily and very fast defined.*

*In general automatic detection of tangency discontinuities is useful for defining the boundaries areas that will be reconstructed separately.*

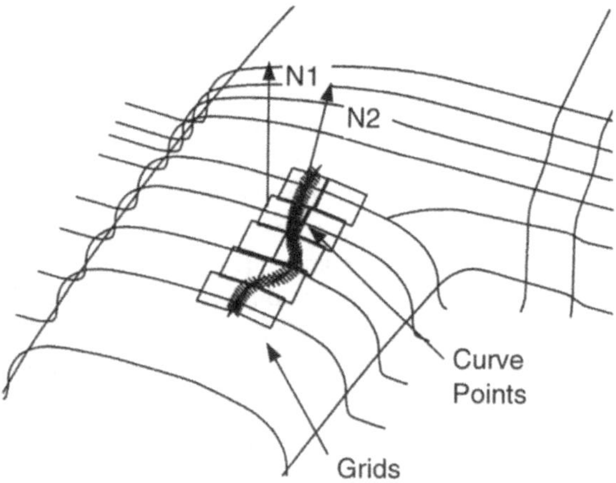


Picture 10: Creation of Boundary contour

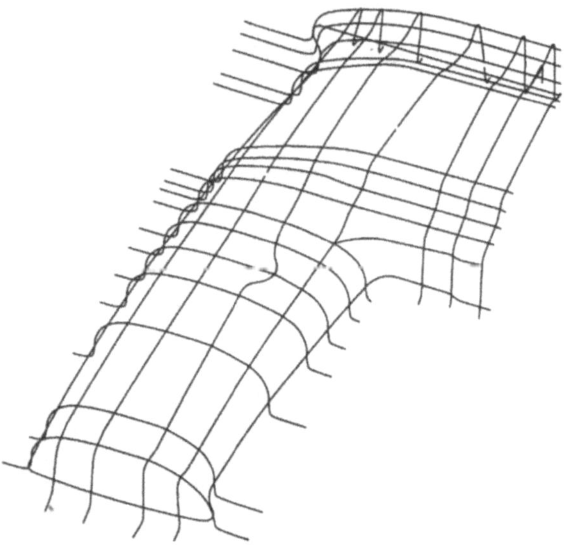

Picture 11: Network

*All these kinds of curves creation are often used because they delimit naturally various areas of the part. The benefit for the user is the powerful assistance for area reconstruction.*

*It has to be regarded that this special functions are supported by standard CAD functionality of STRIM. This is possible because each curve created on base of the grids definition can be handled like any other curve created with STRIM. So each curve can modified by using the wide range of CAD functions if for example the deviation from the digitized points is not small enough and a correction of the curves has to be done interactively. This gives the user very high flexibility to adapt the created curves to his respective wishes and needs. In this context the dynamic adjustability of curves (and surfaces) is a further strong advantage.*

- ***Network Generation***

*When the wireframe skeleton (Picture 11) of the model to reconstruct has been completely defined, the next step consists in the creation of the network. This is done by trimming contours so that they can used as patch boundaries.*

*Automatic functions are available to create single curves between two successive nodes. Network nodes are the points where selected contours intersect. (They may also be the end points of contours). Two contours must have an intersection point within the user defined tolerance. If not they have to be corrected by the user interactivly*

*Furthermore the end points of curves ending at the same node are modified to make them tangent to the same plane.*

*This automatic processes requires contours and curves respectively without gaps. Here the strong basic CAD functions are again a strong help. In case that the automatic contour creation fails and some gaps remain it is an easy work to close or to lengthen them if necessary.*

- ***Surface Generation***

*The network is filled automatically in a single action. Each zone on the network that is delimited by two, three or four boundaries is filled by a patch, whose sides lie precisely on the boundaries. Three types of continuity are possible that can be ensured in the patch sides:*

- *transmission continuity*
- *tangency continuity*
- *tangency continuity with digitizing grid conformity (patch boundaries are tangent to digitizing boundaries)*

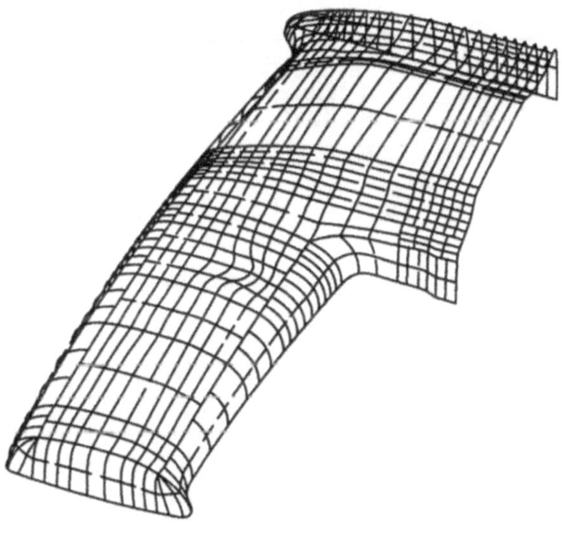

Picture 12: Filling.

*The result of the Surface generation is one single surface (Picture 12).*

Picture 12: Final result of Surface Reconstruction

- ***Surface Control and Modification***

*The result of the Surface Reconstruction process has to be checked concerning different characteristics like tangency, curvature and deviation form the digitized points. For that different capabilities are available.*

*In this context it is again important to have a powerful modelling functionality available. In case of insufficient results mainly the curves and contours of the wireframe skeleton (Picture 11) must be modified with appropriate CAD functions. But also surface wide as well as local optmization of the surface should be possible.*

## *Conclusion*

*Surface Reconstruction is not advantageous from itself. Even if this technology has a wide range of application areas and significant impacts on efficient design and manufacturing important prerequisites have to be fulfilled first.*

*As several time emphasised the availability of strong functions for modelling and quality check are necessary. Furthermore a deep integration of Surface Reconstruction into the basic CAD system is demanded. Otherwise tremendous effort has to be spend to ensure the date exchange between a stand alone Surface Reconstruction application and an independent CAD system. This would lower the general advantages significantly.*

### Author

Dr. Heinz Rybak
MATRA DATAVISION GmbH
Schatzbogen 62
D–81829 München
Tel.: ++49 (089) 42047–0 Fax.: ++49 (089) 42047–100

# Reverse Engineering in Combination with Digital Photogrammetry

*Peter Bonitz, Peter Krzystek*

ICEM Systems Hannover, INPHO Stuttgart

**Abstract:** An online procedure which includes both the digitization of a physical model by means of a photogrammetric measurement system and the CAD based surface description, will be introduced. The CAD system creates class A quality surfaces. With the same procedure, it is possible to update a given CAD model by means of the new raw data (sequences of sample points) taken from a modified physical model. The procedure preserves the structure and parameter distribution of the original surface description in the updated version. The 3D point data from ICEM PHOTO can be obtained in a form suited well for surface construction. The advantage of ICEM PHOTO is that we can avoid the handling of masses of unsorted point data because of filtering. In the last stage of the digital image evaluation we obtain raw data point clouds tailored for surfacing with regard to density and position. The usage of the digital photogrammetric method and the concepts of the methods used for *Reverse Engineering* will be outlined in this paper.

## 1 Introduction

A consistent procedure of digitizing a physical model, describing it in a CAD system, modifying the physical model and redescribing it in the CAD system will be described in this paper. The procedure uses several components. The digitizing is supported by ICEM PHOTO (a digital photogrammetric system) which provides measured data of the physical model. This data is read in to ICEM SURF, a free-form surface modeler for class-A surfaces, where the measured data will be transformed into a mathematical (CAD) description of surfaces with accurate skin lines. In German the term *STRAK-Qualitaet* is used for class-A surfaces. Accurate skin lines are a criterion for highest surface quality in surfaces which have to meet aesthetic requirements, e. g. the exterior surfaces of car bodies. These car body surfaces are verified within CAD systems using curvature diagnosis, reflection lines, and other diagnostic functions. The mathematical model which has been designed with ICEM SURF can be adapted to a modified physical model. During this adaptation the original surface structure and parametrization will remain (almost) unchanged.

The result of digitizing the physical model with ICEM PHOTO is not huge masses of digitized points, but exactly the information needed by the designer to design the CAD surfaces. The digitizing can be automatically applied along desired sections with x = const. or y = const. or z = const., or manually along adhesive tapes. The digitizing can also be performed depending on the curvature. Thus, raw data point sequences are generated which include relatively few points, but are the best possible initial data for the subsequent CAD process. At any time it is possible to create information as detailed as desired because of the existence of the digital photogrammetrical images. The result data of ICEM PHOTO (the ordered point sequences or Bezier patches of order 2 x 2, which are derived from the point data and build up a four sided patch facetting) will be used as input data by ICEM SURF on the same graphics workstation. It is also possible to read the digital images into ICEM SURF as pixel files (background images).

The ordinary iterative way from the raw data to the CAD model using ICEM SURF is also described shortly in this paper. Considering the time it takes to create a class A shape of the whole body, one can understand the wish of designers to keep as much of that detailed geometry description as possible when stylists make subsequent shape modifications. The designers want to adapt the existing surface description to the new digitizing data, which are often measured along profiles. The reverse engineering function of ICEM SURF offers exactly this possibility, if the shape modifications are relatively small in the range of millimeters (as compared to a part of the car body, e. g. a hood). If the shape modifications reach the range of several centimeters, a re-segmentation may be necessary if the curvature characteristics change significantly.

Besides this aim of creating smooth surfaces there is another interesting modeling possibility which helps to save time within the phase of shape development, before the surface description of the final styling model is generated. This is the possibility to create prototype variants within the computer quickly and easily: the facet model created by ICEM PHOTO can be modeled with ICEM SURF just like modeling clay. This can be achieved using the global modeling functions of ICEM SURF.

In the following, the paper outlines in the firts part the photogrammetric approach of ICEM PHOTO. In the second part the *Reverse Engineering* features of ICEM SURF are described and illustrated by examples.

## 2 Digital Photogrammetry

### 2.1 General Remarks

Photogrammetry for a long time used to be well-known as an advantageous measuring technique in various industrial application fields. However, the time-consuming development of the analog photos and the mainly interactive measuring processes reduced the applicability and the acceptance. Since digital images became available by scanners and digital cameras, new industrial application fields have been opened because of the great potential of process automization. Image processing techniques can semi-automate or even fully automate measuring processes which have been slow so far because of the human operator. Furthermore, the instantaneous availability of digital images has also reduced the processing times significantly, so that even real time applications have become feasible.

Digital cameras with large CCD sensors became available only some years ago. Especially, the still video camera DCS200 and the newer version DCS420 from Kodak were equipped with a large CCD sensor with 1500 x 1000 pixel at a pixel resolution of 9 $\mu$m. The newest version of this product line is now represented by the DCS460 which has a size of 3000 x 2000 pixel at the same pixel resolution. This camera enables a photogrammetric system having a high measuring accuracy in a large work area.

As far as the application fields are concerned this paper mainly focusses on the automotive and machine industries where CAx systems are frequently applied. The combination of digital photogrammetry and a suited CAD system is advantageous above all for the iterative process during which the shape and the functionality of a product is optimized. For the sake of completeness let us also mention other application fields like medical technology or the ship building industry. In general it is to be expected that digital photogrammetry will penetrate into a wide application field because of the new flexibility gained by digital cameras.

The first successful application of digital photogrammetry to the measurement of design car bodies was achieved by the InduSURF system [4, 5]. It was installed on an analytical plotter in combination with CCD cameras and used image processing techniques to automatically measure profiles and character lines. The measuring rate of the system of one 3D point per second was, however, rather poor and could not be improved because of the hardware restrictions. This system is today still used successfully at VW and AUDI.

The accuracy requirements for any measurement system vary depending on the application fields. In the measurement of design cars, so far an accuracy of $\pm 0.10$ mm has usually been required, sometimes even in the sense of a tolerance measure, i.e. the standard deviation has to be in the order of about $30 \mu m$. These very tough requirements are mainly influenced by a traditional design philosophy which needs very accurate raw data from the physical model. However, design strategies are changing towards reduced accuracy requirements in the order of 0.5 mm or even 1.0 mm on the full scale model. In this case the design work is shifted to a modern CAD system using visualization tools. Please refer also to chapter 3.4.2. Also virtual design (starting modeling with an empty screen) in the computer is partly replacing the physical design on clay models.

The time requirements for a measurement system to measure a full scale design model are becoming harder since the automotive industry wants to reduce production times as much as possible. Only a few days are acceptable for the design model to be measured. Also, it appears to be important that the object be occupied for a short period only so that the designer can continue his work as soon as possible.

In the following the system approach of ICEM PHOTO will be described in more detail. Special emphasis is laid on the operational steps of the system workflow and on accuracy considerations.

## 2.2 System Approach

The key idea of the system approach of ICEM PHOTO is to capture several digital images of the object of interest which sufficiently overlap the object (see Figure 1).

Digital images are directly recorded by digital cameras (e.g. still video cameras) which can be used like usual photo cameras. Photogrammetric measurement techniques and image processing techniques are combined to digitize the surface of the object automatically. The process produces either dense point clouds or profiles by iteratively adapting the 3D point spacing to the surface curvature. The 3D raw data points are passed to the CAD system ICEM SURF where they are used for the surface description. The system is highly automated with respect to the surface measurement. Interactive input is required for the triangulation of the images to precisely reference the image to the user coordinate system and to properly tie up the images in order to get consistent 3D data in the overlapping areas [1].

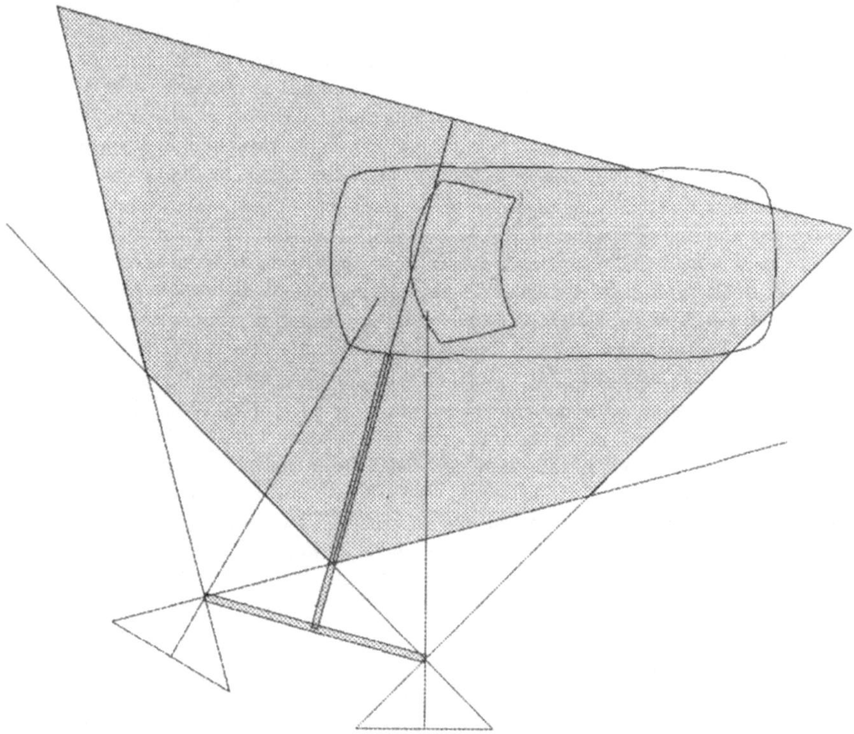

Fig. 1 Principle of stereo photogrammetry.

The hardware of the system comprises two digital still video cameras DCS460 which are mounted on a stereo base. A texture projector is installed between the two cameras to provide a random pattern by using a texture slide. The artificial texture is needed to successfully apply correlation methods in case the object of interest has no natural texture. Figure 2 shows a typical arrangement of the system hardware.

## 2.3 Operational Steps

The operational steps in ICEM PHOTO to be worked through are basically

(1) Camera calibration
(2) Object preparation
(3) Image capture
(4) Data transfer
(5) Photo triangulation
(6) Surface measurement
(7) Output in CAD format

Fig. 2 Camera setup with texture projector.

(8) Editing in the CAD system
(9) Surface description in the CAD system

Since the digital cameras are non-metric they have to be calibrated in a special step in advance. The **camera calibration** (1) comprises parameters for lense distortion (radial and tangential), focal length and principal point. These parameters are derived in a highly redundant bundle adjustment using several images which are taken from a flat calibration point field. The points are defined by black circular dots which can be automatically detected and precisely measured by means of least squares matching. The calibration procedure is highly automated and needs only initial interactive input. Once the calibration parameters have been determined they can be used for a long time. It is recommended to check them periodically within a reasonable time frame of several weeks. The **preparation of the object** (2) comprises mainly the fixation of tags for reference points and tie points on the object (see Figure 2). The 3D coordinates of the reference points have to be measured with a suitable 3D measurement system (e.g. CMM). These points provide the scale and the referencing with respect to the model coordinate system. The tie points are to be measured only in step (5) in the images. They are necessary to accurately tie up the images in the overlapping areas. Eventually, the surface of the object must be tarnished by using for instance an appropriate dulling spray in order to avoid critical reflections in case of shiny object surface. The subsequent **image capture** (3) is operationally quite easy because of the simplicity of the cameras and takes about 2 - 3 hours for half a full scale car design model. Once the images have been taken they are transferred via Laptop (**Data transfer** (4), which is connected on-line to the SCSI disc of the cameras, to the SGI workstation for final processing. Two major steps then have to be worked out. Firstly, the position and the attitude of the cameras with respect to the user coordinate system must be determined. In this **photo triangulation** (5) the reference points and tie points are manually measured in the digital images. Semi- automatic support is provided by correlation techniques. This means that the human operator uses the cursor to provide initial positions for the subsequent refinement based on least squares matching techniques. Multiple image matching is applied, using the classical affine transformation and a consistent parameter setup. In order to promptly get orientation results, the interactive measurement tool incorporates an integrated bundle adjustment package which provides orientation data without initial values and statistical analysis of gross errors based on data snooping techniques. This highly efficient on-line triangulation tool gives maximum precision for image coordinates and reduces the evaluation time for orientation parameters. The result of the photo triangulation provides orientation parameters of the cameras (position and attitude). By using those parameters it is then basically possible to intersect any surface point from at least two cameras. Figure 3 shows the interactive photo triangulation part.

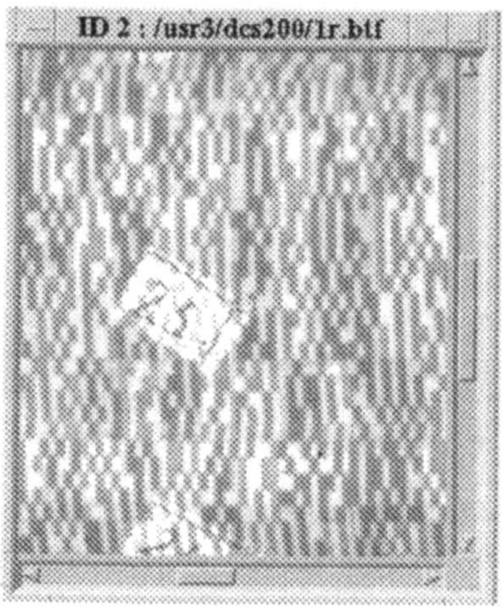


Fig. 3 Reference points to be measured in the photo triangulation.

The photo triangulation (5) is followed by the automatic **surface measurement** (6) which needs some interactive input for preparation. For instance, some starting points, measurement areas and exclusion areas are to be defined. After this preparatory part the automatic surface measurement can be initiated. All individual surface measurements (covering a part of the object to be measured) resulting from the individual stereo pairs can be linked to one consistent raw data set which describes the object surface. It can be **output** (7) in any **CAD exchange format** like VDAFS or DXF. Examplarily, we show raw data of a car body (Figure 4). It is also possible to deliver quadrilateral facets for ICEM SURF.

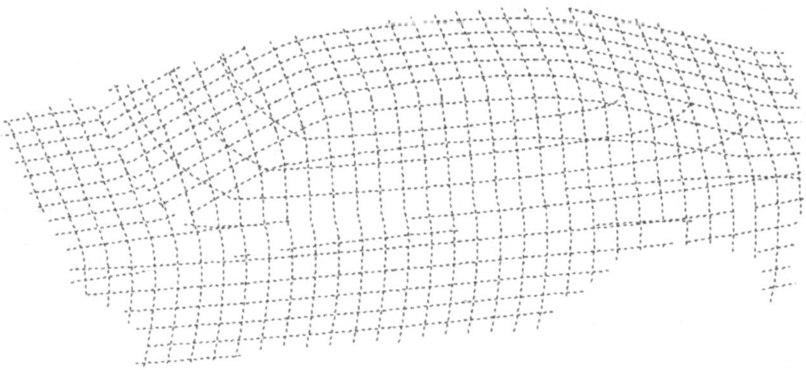

Fig. 4 Raw data point sets for ICEM SURF.

## 2.4 Accuracy Aspects

The system includes basically two measurement options. The first option is mainly based on a topographic approach which has been originally designed for automatic DEM generation [2]. This approach takes advantage of many arbitrarily distributed 3D points which are found by means of feature-based or area-based matching techniques to provide a dense surface description. The points of the cloud are structured in grid lines defining facets or polyhedrons. The orientation of the grid lines is determined by a reference plane which has to be defined in advance.

Basically, the approach is characterized by the feature-based matching technique which is hierarchically applied in image pyramids, and a robust surface reconstruction using finite elements. Measured 3D points, together with curvature and torsion constraints are introduced as observations. Since the observations are weighted down in a two-step iterative least squares adjustment, the algorithm is principally suited to preserve surface edges like character lines. The spacing of the grid points is by default relatively small (e. g. 15 pixels) and is automatically adapted to the local surface curvature. Because of the hierarchical approach through image pyramids no initial surface is required to start the surface reconstruction.

Especially in the automotive industry, the digitization of design models is required along profiles which are defined by the intersection of pre-defined planes with the surface. The intersection planes are usually parallel to the model coordinate system. In some cases special planes need to be defined perpendicular to 3D polylines or rotating along a straight line as so-called radial planes.

The measuring strategy in this second option incorporates hierarchically applied area-based and feature-based matching and guarantees that the surface is measured with profiles wherever possible. Problem areas caused by poor texture or reflections, in which matching techniques fail, are circumvented in order to keep gaps as small as possible. Only one starting point has to be provided in advance, whose digitization accuracy must fall in the pull-in range of the matching algorithms. Also, the point density in the profiles is automatically adapted to the surface curvature by forcing the measured 3D points strictly into a tolerance hull.

For a short accuracy consideration, let us presume a typical stereo camera configuration with a camera-to-object distance d = 1.3 m and a base-to-height ratio b/d of 1/2 (see Figure 1). The camera in use should be a DCS460 digital camera with a focal length of 24 mm. The pixel size of the digital image should be 9 $\mu$m meaning that the pixel size on the object is in the order of 0.6 mm.

Since the matched feature points of the point cloud generator (first measurement option) have a precision of 1/3 pixel, the internal measuring accuracy of a single 3D point in object space is expected to be 1/3 pixel *60*2 = 0.5 mm. If the object is for instance covered by an artificially created texture (e. g. projected random pattern), typically 30 to 40 measured 3D points are found in a single finite element mesh with a size of 10 pixel x 10 pixel = 6 mm x 6 mm. Because of this high redundancy, the points of the finite element get a higher accuracy than the single matched 3D points, which is for this particular case in the order of 0.16 mm. Nevertheless, the individual matched 3D points can easily be refined by least squares matching techniques, which is reasonable in case of smaller finite elements or in the presence of poor texture. With regard to the camera configuration mentioned above, a point cloud covering an object area of about 1 m x 1 m with approx. 30,000 - 40,000 3D points on average is generated.

The internal measuring accuracy of the profile measurement option is directly influenced by the inherent potential of the area-based matching technique being applied by default. If we recall the camera given above and presume a parallax precision of 1/10 pixel, the standard deviation of a single profile point amounts to 0.1 pixel * 60 *2 = 0.11 mm. Actually, this is in the same order of magnitude than the point cloud precision.

## 2.5 System Performance

Finally we should mention figures reflecting the system performance. The measuring speed is mainly dependent on the workstation used and — in case of an Indigo2 with an R4000 MIPS processor — amounts for the point cloud generator to 50 3D points/sec, and for the profile measurement option to 25 - 30 3D points/sec. This nominal measuring rate will be improved significantly as soon as the next computer generation becomes available.

More important for the practical acceptance is the time needed for the complete measurement of e. g. half a full scale design model. Half such an object could be covered by about 15 - 20 stereo models (= 30 - 40 images) presuming a camera-to-object distance of about 1.3 m and a base-to-height ratio of 1/2. Altogether, the steps described in chapter 2.3 may take the following time:

| Step | time required |
|---|---|
| Object preparation | 2 h |
| Image capture | 3 h |
| Data transfer | 3 h |
| Photo triangulation | 5 h |
| Surface measurement | 8 h |
| Data editing | 3 h |
| Total: | 24 h |

The average point spacing would be in the order of 5 mm on the object. In areas of large surface curvature it would amount to about 1 mm. Thus, about 30,000 or even more 3D points would cover the object. At this point the advantages of digital photogrammetry become quite evident. The user can feel free and can measure the object with a point density which is optimal for the subsequent CAD-based surface description. There is almost no limitation to the number of 3D points, unless the point spacing is not significantly smaller than the pixel size. In practice, it will be an iterative process between measuring system and the CAD-system in which the best point density depending on the complexity of the surface is found. Actually, the redigitizing of certain areas of the design model can be accomplished be solely using the images. The object is not needed since the images contain all the information for remeasurement. Furthermore, it is also worthwhile mentioning that the raw data the measuring system provides is filtered data which is automatically adapted to the surface curvature.

## 2.6 Benefits

Let us summarize the advantages of the system approach.

- ICEM PHOTO is a handy and easily transportable measurement system. By using the hardware component, the design model of interest can be quickly captured by a couple of digital photographs in-house or in other design studios. The digital images can be sent on the Internet to different places for the subsequent automatic surface reconstruction.

- By using ICEM PHOTO the design model is only occupied for the short period of taking the images. Thus, the designer can continue his work very soon.

- The digital images document each step in the design process. Particular areas of the design model can be re measured by ICEM PHOTO without using the design model which might be in use in the current design work.

- ICEM PHOTO takes advantage of highly sophisticated calibration techniques which are applied to the digital still video cameras. The accuracy is in the order of some tenths of a millimeter depending directly on the distance between the camera and the object. The measurement rate of the automatic surface reconstruction is in the order of 10 - 20 3D points per second. Half a full scale model may be digitized in 3 - 4 days including all process steps like object preparation, image capture, image orientation, and surface reconstruction.

- The design model is digitized either classically in profiles or as a structured dense point cloud. Robust statistical techniques clean the measurements from gross errors and filter the digitized data.

- The cameras can be flexibly arranged in any position. Thus, certain areas of the design model can be digitized for special purpose.

- The digitized surface data is provided in the model coordinate system. Special photogrammetric techniques guarantee to accurately tie up the single surface parts.

- ICEM PHOTO is integrated into ICEM SURF. Special data output formats are provided to use ICEM SURF's visualization techniques.

# 3 Reverse Engineering

## 3.1 Terms and Phrases

The English term *Reverse Engineering* and the German term *Flaechenrueckfuehrung* have become popular in the CAD-using industry. In Germany the term *Flaechenrueckfuehrung* originates from the process where CAD models needed to be modified according to manual shape modifications to the physical model. The CAD model had to be modified to re-establish conformity of the physical and mathematical models. This task, i. e. to adapt the original CAD model to newly digitized data of the modified physical model, was referred to as *Flaechenrueckfuehrung* in German. It can also be described as the correction of the original mathematical surface description according to newly created raw data point sequences (measured along profiles in most cases).

After some time the meaning of this term *Flaechenrueckfuehrung* was enhanced. Now it also covers the meaning of the English term *Reverse Engineering*. This term means that a CAD model description is derived from a digitized physical model.

It would be much easier and clearer to call both tasks *surface description (based on raw data point sequences) with (— task 1 —) and without (— task 2 —) a CAD reference model.* The aim of the surface description may differ depending on the field of application. The measured or digitized data can be provided as an unsorted point cloud or along sections/profiles.

In the following chapters we will discuss these two tasks and their solutions by means of simple examples. In these examples the raw data used was not obtained in any case by photogrammetry. The intention, however, is to explain the basic ideas of the *Reverse Engineering* methods in ICEM SURF. These methods can be considered as independent of the measuring method. A CAD system should be able to transform regularly and irregularly distributed raw data point sets into a surface description. Especially the so-called *Suction Method* meets this requirement.

This method has been designed specifically for task 1 (raw data with a CAD reference model). Task 2 (raw data without a CAD reference model) is quite an ordinary job in ICEM SURF and from our point of view does not have anything in common with *Flaechenrueckfuehrung* in its original sense. However, the user can also apply the *Suction Method* in case of task 2. It is an additional tool for creating spatial shapes. The user can test these new possibilities and find out the best way for his particular requirements. The best solution is probably a combination of the functionality the user is familiar with, and the new features of the CAD-system.

## 3.2 Combination with ICEM PHOTO

Because we are convinced of the advantages of digital photogrammetry (please refer to chapter 2.6) we prefer this way of obtaining our raw data to other measuring methods. We are currently in the process of gaining experience in how to handle this photogrammetric data. The first results are very satisfying.

## 3.3 CAD Reference Model Exists

As an example we want to look at the outer surface of a door. See Figures 5 to 10.

### 3.3.1 Task

The CAD model and the measured data of the geometric object are given. The sample points have been derived from the modified physical model. They have already been adjusted to the CAD model, see Figure 5. The crowning of the outer surface of the door was enforced. The newly digitized data is measured along x = const. profiles. It is displayed as dashed lines. Figure 6 shows the x = const. sections as well as the digitized data having the same x-values. The deviations between the sections of the original surface and the newly digitized data are shown with a scaling factor of 10 in Figure 7. The adapted CAD model is requested. The CAD model is to be adapted to the new raw data, preserving the surface structure as far as possible.

### 3.3.2 Solution: Difference Surface

*Suction Method*

Figure 9 shows the deviations after the door has been modified by the reverse engineering function using the same scaling factor. The final deviation is clearly visible in Figure 10. The original patches are *sucked* to the digitized data, preserving the surface structure as far as possible. The *suction principle* is very simple. Patch by patch the given CAD model and the digitized data determine a difference surface. The difference surfaces DS will be added to the original surfaces OS (see Figure 8). Thus, the resulting surfaces RS will be created: RS = OS + DS. The resulting surfaces RS will be computed such that the parametrization of the original surface is preserved as far as possible. A satisfying result will be obtained only if the differences are relatively small and if the newly digitized data is distributed almost regularly across the original patches. If the original surface is composed of Bezier patches with continuous transitions ($G^1$ or $G^2$ transitions), the addition may violate the conditions. However, provided that the differences are small, the violations will be nearly invisible. If necessary, the continuous transitions can be restored using the appropriate functions of ICEM SURF. Nevertheless, adding the difference surfaces also works with large differences. It will, however, require manual corrections to close the resulting gaps and to restore continuity at the transitions. If you transform the Bezier patches into a B-Spline surface in advance, the continuous transitions will be preserved.

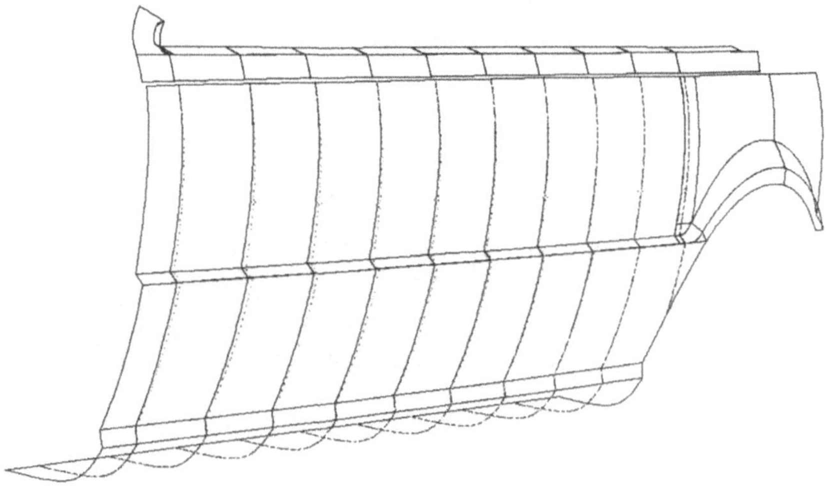

Fig. 5 Zoom window of the CAD model, original patches to be modified, and new raw data. State before using the reverse engineering function.

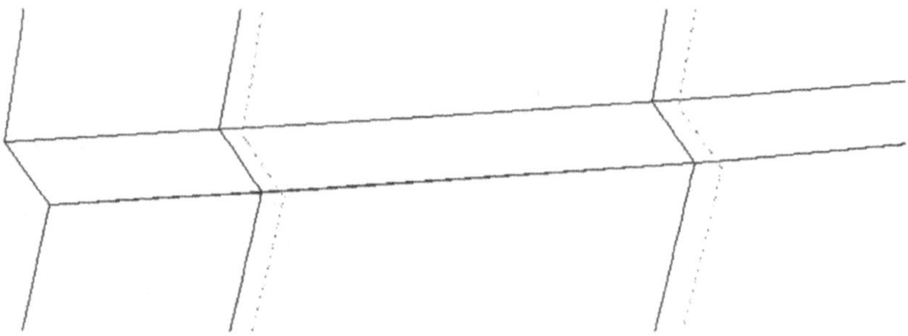

Fig. 6 Enlarged zoom window of the CAD model. State before using the reverse engineering function.

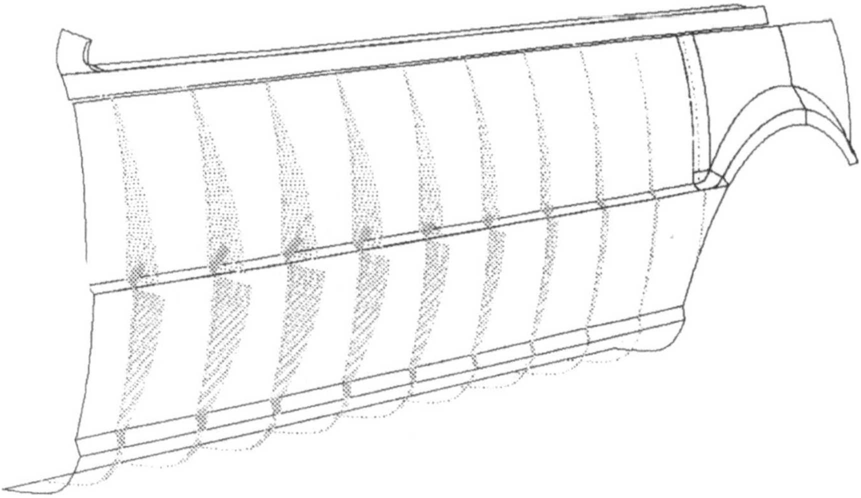

Fig. 7 Diagnosis of the deviations. State before using the reverse engineering function.

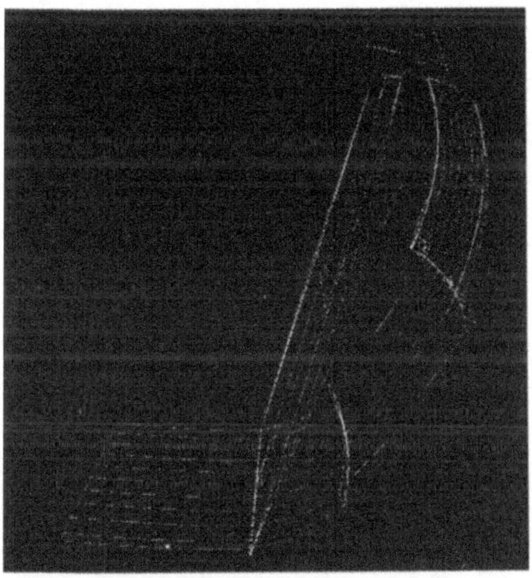

Fig. 8 The difference surface is displayed. The control points of the difference surface can be interactively adjusted.

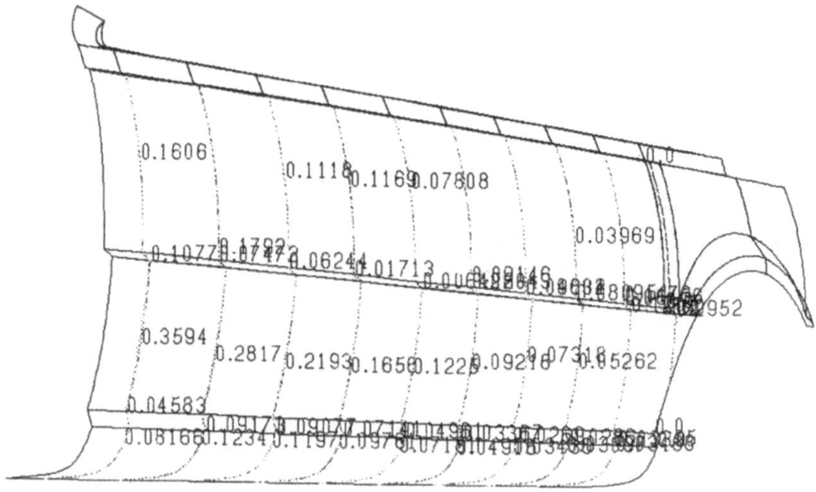


Fig. 9 Diagnosis of the deviations. State after using the reverse engineering function.

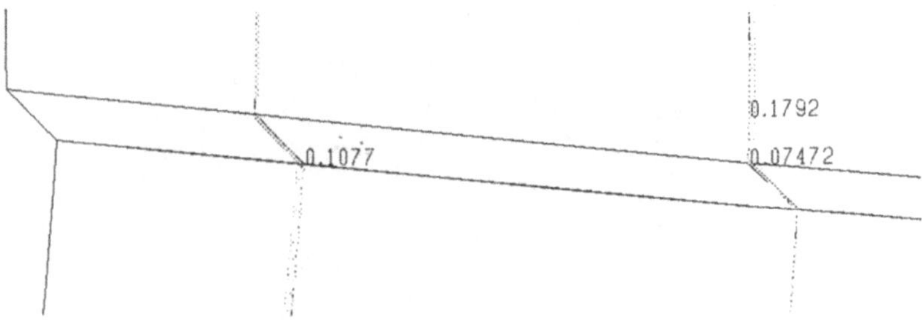


Fig. 10 Enlarged zoom window. State after using the reverse engineering function.

### 3.3.3 Integration of the Measured Data into the CAD Model

Adopting the raw data from ICEM Photo bears the advantage that the sample points include an adjusted 3D position to the definite coordinate systems. If the manual way of digitizing (e. g. via 3D - coordinate measuring machines) is chosen this is not guaranteed. However, in this case the modified physical model can optionally be rotated and translated against the original coordinate system via the moving functions of ICEM SURF. This can be carried out manually for example by using the associative diagnostic functions. You may wish to observe the distance diagnosis between CAD model and digitized data while rotating or translating the digitized data. After a few iterative steps the position is found where the deviations are smallest. Alternatively you may define three (or more, in which case you need a least square fit) connection points.

## 3.4 CAD Reference Model does not Exist

### 3.4.1 Task

Only the measured data of the geometric object is given. The CAD surface description is requested.

### 3.4.2 Ordinary Solution: Interactive Associative Surface Approximation or Construction

In the following we will describe the quite normal daily work of an ICEM SURF user. He is familiar with raw data and the errors they usually include. We mentioned above that the absolute accuracy of the digitized points can reach $\pm 1/10$ to $3/10$ mm, if we use KODAK DCS 460 cameras with 3000 x 2000 pixels. The accuracy also depends on the camera position with respect to the measured object. Please refer also to chapter 2.4. Randomly occuring errors do not seriously influence the surface description process. It is not important to create an exact interpolation through each measured point, but to recognize the global trend of the curvature within user-defined areas. As trend surfaces we usually use Bezier patches of the lowest possible order. These patches can be created by approximation (least square fit) of the point cloud after the user specified the edge curves. The algorithm used also takes into account the measured erroneous data between the four edges. Experienced users, however, prefer the interactive bending of initially planar patches to the mentioned approximation. The bending interactions for creating the desired spatial shape comprise definite movements of single or groups of connected control points with degree elevation. The *Deviation Diagnosis* between patches and measured data will visualize a *random noise* of the measured data of approximately 0.1 to 0.3 mm, if you use the camera type mentioned above and a similar camera setup mentioned above in chapter 2.4. The process of describing a surface based on raw data point sequences with ICEM SURF will be successful if you keep in mind the following rules and principles: Please create a minimum number of patches having the lowest possible order and a segmentation which is exactly suited to the curvature characteristics of the whole geometry. This means that the patches should be as large as possible, and every patch in itself should have rather constant curvature values. Please create patch boundaries, if the curvature changes from flat (low curvature) to bent (high curvature). Please make use of all associative diagnostic functions to evaluate the shape of the created patches and the geometry as a whole. The process of creating surfaces from raw data with ICEM SURF avoids the time-consuming and difficult iterative process of creating curves from raw data and surfaces from curves. Using ICEM SURF you may directly create patches from raw data. These patches can be modified using the control point modification functions and evaluated by the just mentioned diagnostic functions, leading to the initially mentioned accurate skin lines.

### 3.4.3 Additional Solution: Difference Surfaces

The principle of adding a difference surface can also be used for a mathematical description without having the CAD reference model, see Figures 11 to 13. First, planar patches need to be designed in a suitable plane. The experienced designer will recognize the most suitable subdivision of the surface according to the rules presented in chapter 3.4.2. He will only determine the planar edge curves of the patches. Creating the planar patches from these edge curves is a routine task which can be performed automatically. The distances between the single points of the raw data point sequences (digitized data) and the planar patches in the direction of the plane normal will be computed to a total difference surface. Figure 11 shows the state before execution of the reverse engineering function, whereas Figure 12 shows the state after reverse engineering has been performed. The result is a rough surface description which can be modified very easily using the different modelling functions of ICEM SURF. After manual corrections of the edge curves and repeated *suction* steps one can achieve the final result as shown in Figure 13.

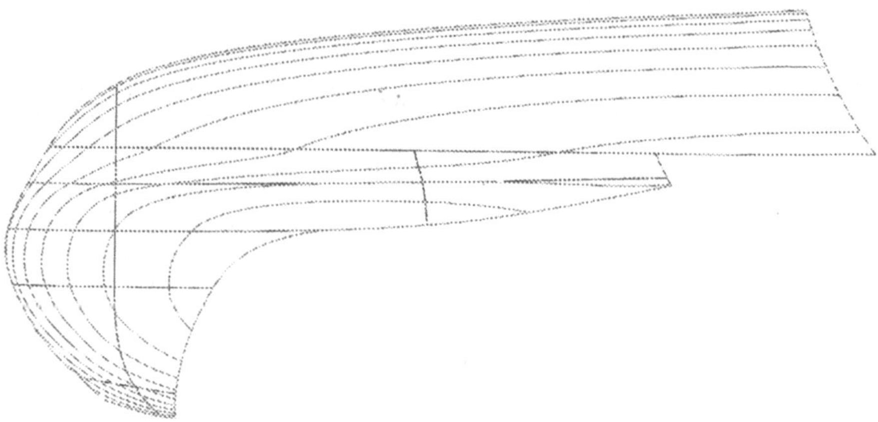

Fig. 11 Planar patches and measured raw data. State before using the reverse engineering function.

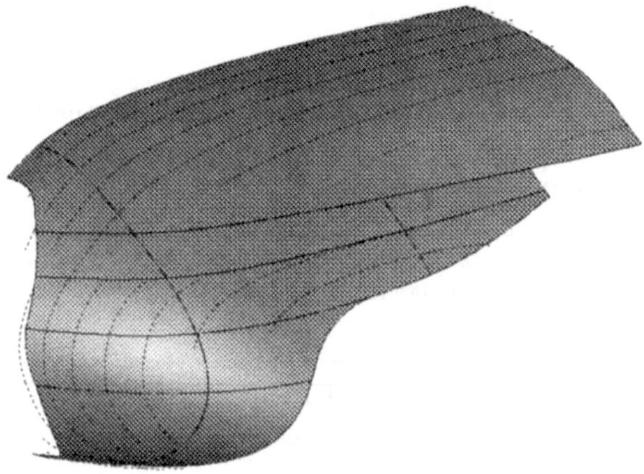

Fig. 12 Result of the *suction method* after the first iterative step.

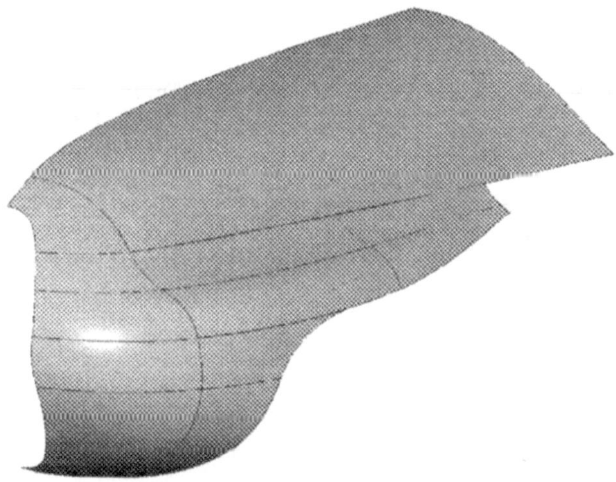

Fig. 13 Final state after manual correction.

# References

[1] Gruendig, L., Buehler, W.: Zur Naeherungswertbestimmung und Buendelausgleichung von Konvergentaufnahmen. Bildmessung und Luftbildwesen 53 (6) 1985

[2] Krzystek, P.: Fully automatic measurement of digital elevation models, Proceedings of the 43rd Photogrammetric Week, Stuttgart, pp. 203 - 214, 1991

[3] Krzystek, P., Petran, F., Schewe, H.: Automatic Reconstruction of Concept Models by using a Digital Photogrammetric Measurement System, IAPRS, Vol.30, Part 5W1, ISPRS Intercommission Workshop, From Pixels to Sequences, Zurich, March 22-24, 1995

[4] Schewe, H.: Automatic Photogrammetric Car-Body Measurement. Proceedings of the 41st Photogrammetric Week, Stuttgart, pp. 47 - 56, 1987

[5] Sorgatz, U.: Das Geometriedatenerfassungssystem VWSCAN im CAE-Konzept, VDI-Z 7/88, pp. 38-46, 1988

Addresses of the authors:

Dr.-Ing. habil. Peter Bonitz
Kuesterstr. 8
ICEM Systems GmbH
D - 30 519 Hannover
Phone:(+49) 511 - 98 48 - 833
Fax: (+49) 511 - 98 48 - 810
E-mail: pb@icem.de

Dr.-Ing. Peter Krzystek
Smaragdweg 1
INPHO GmbH
D - 70 174 Stuttgart
Phone: (+49) 711 - 22 88 - 120
Fax: (+49) 711 - 22 88 - 111
E-mail: peter@inpho.de

# An Introduction to Sculptured Surfaces Metrology (Holometry)

*Carlos G. Porta, Roland Fröwis, Udo Kleindienst*

Holometric Technologies GmbH

## 1. Introduction

The title of this workshop "***Reverse Engineering***" is to demonstrate that this meeting is above all concerned with "***surface reconstruction or mathematical modelling using 3D points***". Nevertheless the authors are of the opinion that this is only a link in the chain of a more general task, which is : **Manufacturing Technology.** In this context, the aim of Sculptured Surfaces Metrology using Coordinate Measuring Machines and the HOLOS software package, specially developed for this application field, will be introduced.

*"A machine without software seems like an automobile without wheels"* is a popular saying in modern industry . It is very well known that software is becoming more important than hardware -more important than machine tools- in factories all around the world. Call it "*the digital, or soft or virtual factory*" for it depends on information technology .
The experience of the last 15 years in developing software for industrial applications ( after the introduction of CAD technologies) shows very clearly that the goal "*the universal software package* " is far away from the reality. One of the strongest limitations might be, is the human factor; his particular way of thinking depending on his type of education. The authors of this article are of the opinion that a good compromise solution is to use "**problem oriented software packages**" having clear and flexible interfaces to the general task mentioned above.

## 2. Manufacturing Technology

***What are the permanent principal objectives of manufacturing technology ?***

- To increase the quality of products and production processes as a result of improved machining and measurement accuracy
- More economic efficiency due to flexible automation and powerful software
- Shorter production times

Increasing the quality of a product not only concerns the quality of the materials. Due to the mechanical interaction of different parts it is necessary when machining to take into account very close tolerances.

***What does this mean in the field of dimensional and geometrical metrology ?***

To develop the metrological basis for accurate measuring instruments and their reliable application in quality control loops of modern production systems. In a few words the main objective of geometrical metrology is :

*"To achieve early information on the geometric parameters : dimension, form and position, to allow the machining process to be corrected in good time so that reworking and rejects are avoided and finally so that* ***quality can be kept under control"***. ( See [2])

## 3. Coordinate Measuring Machines

The past fifteen years have seen a continued increase in the importance of Computer Aided Measurement Techniques as a means to control modern production engineering. Coordinate Measuring Machines are inspection units which can be used not only for accurate measurements of workpiece dimensions in any direction, but also for the reliable measurement of the shape and position of almost any complex workpiece feature.
A Coordinate Measuring Machine in brief, a "transversing frame" with three orthogonal axes (See Fig 1).

*"Boss,I have the feeling that we are exaggerating a bit, aren't we ?"*

*Fig. 1: Schematic representation of a Coordinate Measuring Machine.*

These axes, designated X, Y and Z, embody a spatial co-ordinate system, similar to that of an NC milling machine. The displacements along the axes are read by length-measuring systems and transferred to an electronic control cabinet. In very simple, manually operated machines they are displayed directly. (See [1]).

## 4. The HOLOS software package

**HOLOS (from the Greek : whole, entire) is a state-of-the-art software package for the purpose of measuring and digitising sculptured surfaces together with a CMM.**

***Measuring*** : which means capturing the form deviations between the generated sculptured shapes (nominal geometry) and the machined workpieces (actual geometry).

***Digitising*** : which means transforming a physical model into 3D points for which no surface description exists, allowing them to be processed further in a CAD System.

HOLOS is based on the following configuration :

- The UNIX Operating System
- The X-Window User Interface

HOLOS features X-Windows, providing a highly convenient user interface for which no specific CAD background is required. It was developed and optimised in co-operation with our customers and incorporates optimum software ergonometry. This results in a logical structure which reflects the way of thinking of a CMM operator. The HOLOS user interface conforms to OSF/Motif guidelines. All functions are activated by pulldown menus or buttons, thus minimising input errors.

*Visualisation* is a vital factor in defining the different measuring tasks. HOLOS offers the convenience of a graphic interactive system with the functions *Zoom, Translation, Rotation and 2D Projection.* Spatial identification and object selection is made very easy by just clicking the mouse.( See Fig. 2 and Fig. 3)

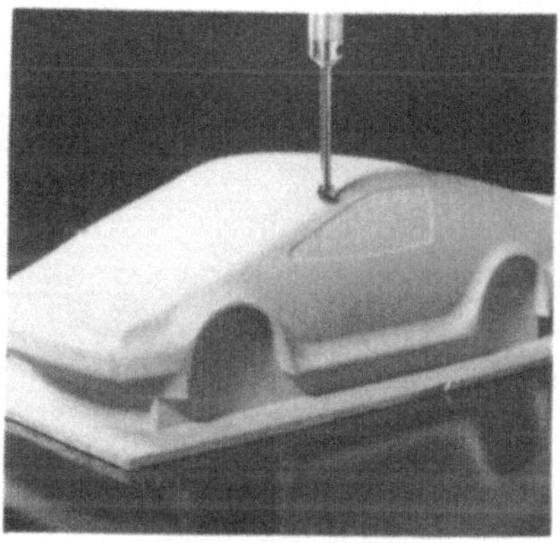

*Fig. 2 : Digitising a car model.*

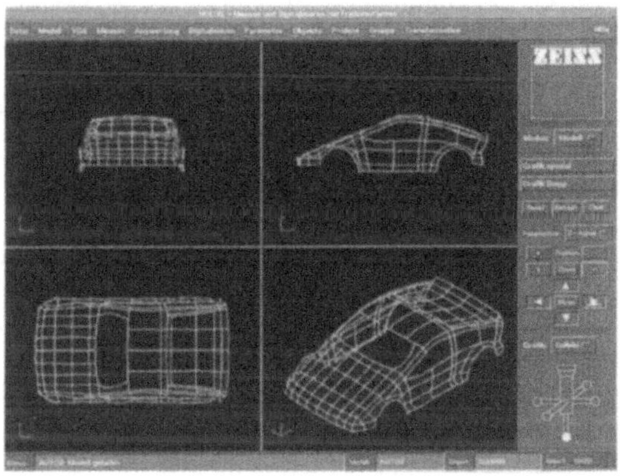

*Fig. 3: Graphical representation of the car model and HOLOS User Interface*

*Geometry data interchange* with other CAD systems is performed using the VDA or IGES interfaces.

### 4.1 Different Strategies for Digitising a Physical Model

Digitisation of models is not a trivial task because the choice of the strategy or method depends heavily on the field of application.

***What is the meaning of "accuracy " in a digitisation and re-engineering process?***

This is a term whose meaning depends on the end-user application. The following main fields of application have to be differentiated:

- **Copy milling** : The acquisition of a large number of points on the model surface facilitates direct linking to a CAM module to obtain NC data for controlling the machine tool. In this case only the dimensional accuracy is required and is derived from the accuracy of the measuring device (CMM or optical devices). Mostly, the strategy is to define different measuring areas, and each one is scanned in scanning lines (or 2D Sections).

- **Reverse Engineering** : In this field, the objective is to gain the surface description of the model. The final accuracy is a combination of different components:

***Dimensional accuracy*** : which measures the "goodness of fit" of the digitised points to the surfaces obtained. Dimensional accuracy is very important in metrological applications. A typical situation is when a "master model" is digitised and reverse engineered to use its surface description to measure additional parts in a production line. A very good example is the geometrical control of gears and also aspherical optical surfaces. In general, as in this case, the geometrical accuracy is of second order.

***Geometrical accuracy*** : defined by the intrinsic characteristic of the fitting surfaces, such as the kind of continuity (C(0), C(1) or C(2)) within the different surfaces, and the smoothness. Geometrical accuracy plays a crucial role in the field of Visualisation and Design due to the requirements of aesthetic profiles, such as car bodies and household goods. In such cases the continuity conditions within different surfaces have to be guaranteed, as well as a good smoothness behaviour.

- **Finite Elements Analysis**

Here point distribution is more relevant than geometrical and dimensional accuracy.

Now, in order to clarify these ideas, the digitisation of a car wing using a CMM equipped with a laser sensor will be described.

The steps are :

- **Defining the boundary curves of an area.**
- **Initial surface approximation using "lofting".**
- **Defining the scanning parameters and automatic scanning.**
- **Surface Approximation**

### 4.1.1 Defining the boundary curves of a area.

This is a manual procedure and consists of recording several points approximating a curve segment. The number of points must be proportional to the complexity of the curve itself. (See Fig. 4).

### 4.1.2 Initial surface approximation using "lofting".

After this, using the **lofting** function, a rough approximation of the area to be scanned is obtained. A rough approximation but with enough information to define the control parameters of the CMM for scanning the lines, that is : the start and end points and a "pseudo" normal direction to control the laser head. (See Fig. 5)

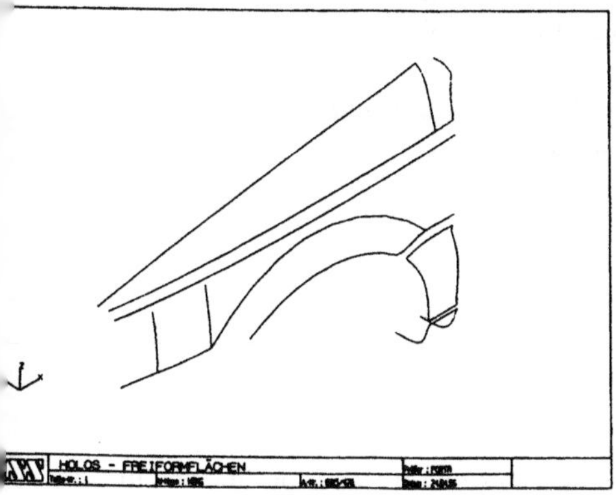


*Fig. 4: Characteristic curves of a car wing*

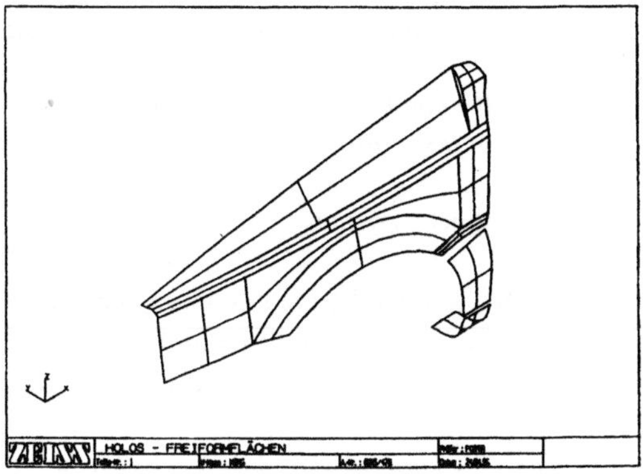


*Fig. 5: Initial surface approximation*

**4.1.3 Defining the scanning parameters and automatic scanning.**

Before starting the automatic scanning (see Fig. 6) it is necessary to define the control parameters, which are:

- **Speed of the measuring machine**
- **Distance between two scanning lines**
- **Curvature and step by step tolerances if any filter is desired to decrease the number of points.**

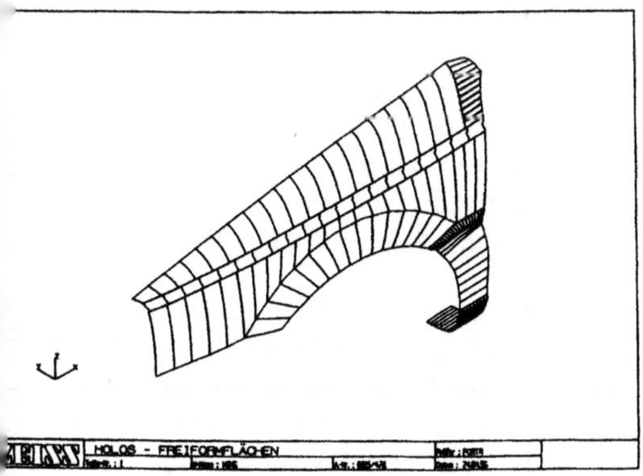


*Fig. 6: Scanning curves*

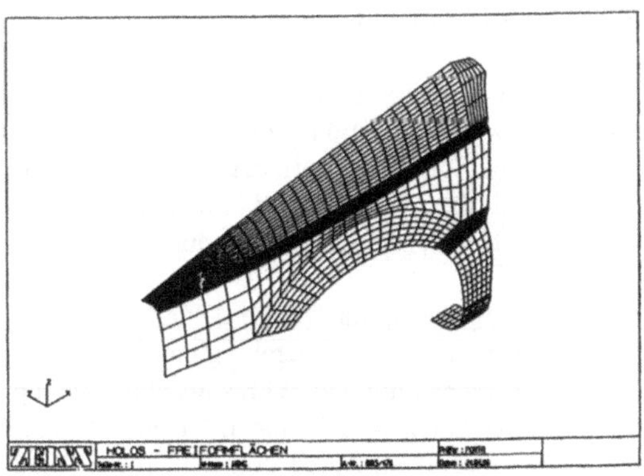


*Fig. 7: Surface Approximation*

#### 4.1.4 Surface Approximation

After scanning the area, it is possible (not necessary) in HOLOS to improve a surface approximation (see Fig. 7) using the ***Least Squares Method***, which is:

***"Find the coefficients in the B-Spline representation so that the sum of the squares of the distances taken in orthogonal direction, is minimised".***

The parameters required are:

- **Polynomial degree in u and v directions**
- **Number of patches in u and v directions**
- **Continuity condition : C(0), C(1), C(2)**

One of the greatest advantages of having a system like HOLOS, which is on-line and directly connected with a CMM is the possibility of making an accuracy control immediately after the digitisation. If something goes wrong then the operator can introduce the necessary changes in his strategy.

Following the same procedure for every scanning area, the surface description of the whole car wing is then obtained. (See Fig. 7 )

### 4.2 Different Strategies for Measuring a Physical Model

As explained before, to measure a model means: ***to capture the form deviations between the sculptured shapes generated (nominal geometry) and the machined workpieces (actual geometry).***

In HOLOS there are two measuring methods : Manual and automatic mode.

#### 4.2.1 Manual Operation

This is the most simple operation mode. Here, the special advantage lies in the extreme speed at which deviations are determined and illustrated, even on workpieces described by a very large number of surfaces. Regardless of the task, single point probing can be performed at will; the surface gained is then automatically recognised, the probing points projected on the surface in orthogonal form and the deviation recorded as both a graph and a measuring record.

Using the colours of the deviation flags, the user can decide whether the result is within the tolerance or not, whereby green displays results within tolerance, blue results below the lower tolerance limit and red results above the upper tolerance limit. In addition to this information, a deviation distribution is displayed in a separate window and here the user can view the individual measured points in more detail (X, Y, Z deviation and deviation in the normal direction), he/she can also delete measured points.
These points can be stored if the manually measured points are to be measured e.g. on another part or if they are to be archived. In this way a CNC program is created without any more work needed.

To sum up it can be said that the manual method offers a simple, fast way of measuring and programming a CNC run in the 'teach-in mode'.

### 4.2.2 Automatic Mode

The next step is the generation of measured points interactively on the screen using graphics. HOLOS offers the following strategies:

- **Plane Section**
- **Grid in parametric co-ordinates**
- **Grating in Cartesian Co-ordinates**
- **Single Points: defined in HOLOS**
- **Parallel curve (Offset Curve)**
- **Single points imported from a CAD system**
- **Scanning line**
- **Scanning area**

#### 4.2.2.1 Plane Section

Using the Plane Section, the start and target points are defined in the area desired using the mouse. The system then generates the number of points entered distributed evenly along this line or depending on the curve a calculated number of points. (See Fig. 8).

#### 4.2.2.2 Grid in parametric co-ordinates

Using the grid, the number of points to be generated is input in two directions (e.g. 5x5). These points are then distributed evenly on one or several selected surfaces. (See Fig. 9).

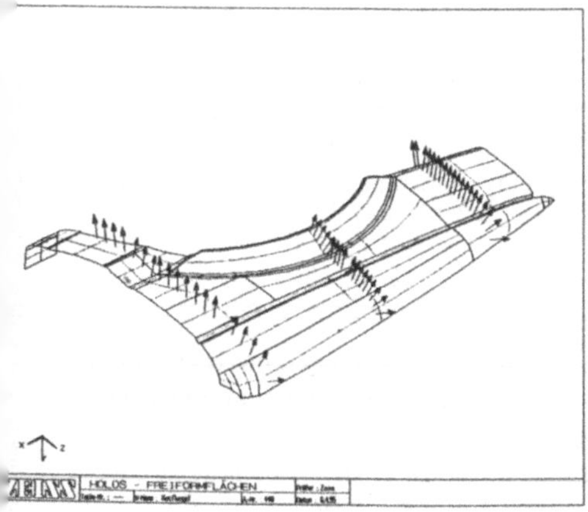

*Fig. 8: Measuring points on plane sections.*

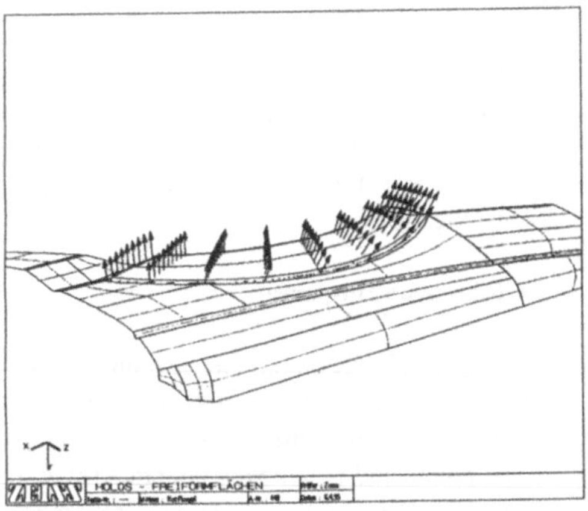

*Fig. 9: Measuring points on a regular grid.*

#### 4.2.2.3 Grating in Cartesian Co-ordinates

When generating grating points, the four corner points of the grating are given using the mouse which has the advantage that the grating can be defined over several surface elements. Also in contrast to the grid, a curvature-dependent point distribution is possible. (See Fig. 10).

#### 4.2.2.4 Single Points: defined in HOLOS

Another possibility is to define single points on the screen using the mouse. The co-ordinates of the points in question are displayed in a separate window. (See Fig. 11).

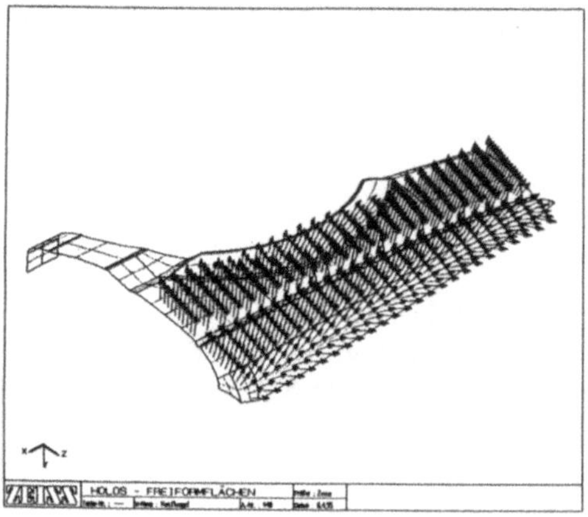

*Fig. 10: Measuring points on a grating*

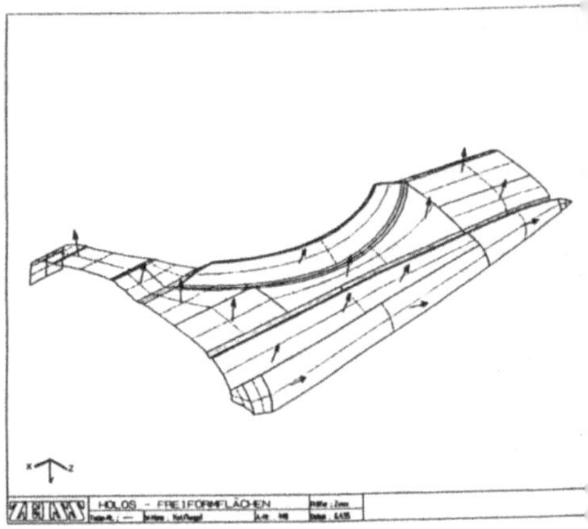

*Fig. 11: Single measuring points.*

#### 4.2.2.5 Parallel curve

An important function for the measured point generation is the parallel curve as in many cases it is not possible to measure up to the edge of a surface / face boundary or a defined distance to such an edge is required. To generate the parallel curve, the corresponding element (surface or face) is selected and the distance to the boundary curve given. Measured points are then generated on the parallel curve which are either evenly distributed or curvature-dependent. (See Fig. 12).

#### 4.2.2.6 Scanning area

The second option is the scanning area which is limited by a polygon line similar to digitisation. In this case, a projection plane as well as the distance to the scanning lines have to be entered in addition to the technological parameters.
The points and parameters generated with the various methods are then stored as nominals or parameter files and can then be started at any time as CNC run. (See Fig. 13).

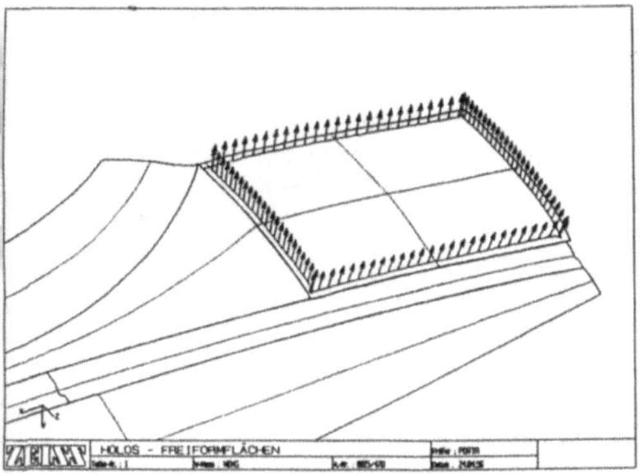


Fig. 12: *Measuring points on a parallel curve to the boundary curve of a surface.*

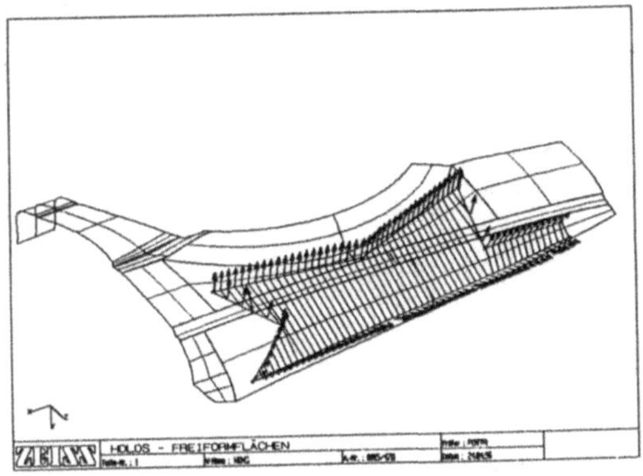


Fig. 13: *Scanning a zone*

### 4.2.2.7 Points imported from a CAD System

CAD points offer a further option for generating measured points. With this option, points defined in a CAD system are read in HOLOS and displayed. These points can then be selected with the mouse and stored as measuring run.

### 4.2.2.8 Scanning line

All the methods described till now are restricted to point-to-point measurement. In the following, two methods are shown for continuous point recording (scanning).

The start and target point are defined on the screen with the mouse and then technological parameters such as step size, scanning speed, angle change for curvature dependency, etc. are entered.

## 4.3 Representation of the deviations

In the next section, the various ways of displaying deviations graphically will be shows. It should be mentioned that the type of display depends on the number of measured points.

The following types of representation are possible in HOLOS:

**4.3.1 Deviation flags** : This is a good way of showing individual deviations using colours (within tolerance, below the lower tolerance, above the upper tolerance) and the actual numerical value of the position measured. (See Fig. 14).

**4.3.2 Vectorial representation**: With this type of representation, the deviation can be displayed in the form of vectors which again contain colour information (in, out of tolerance), as well as supplying information on the amount of deviation using the size of the vector. (See Fig. 15).

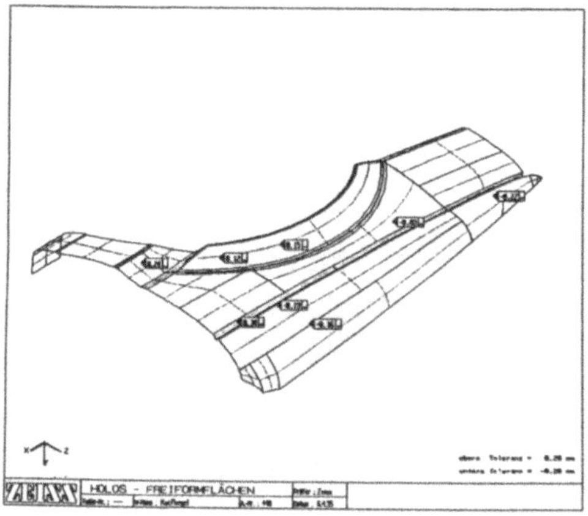


*Fig. 14: Deviations represented as flags.*

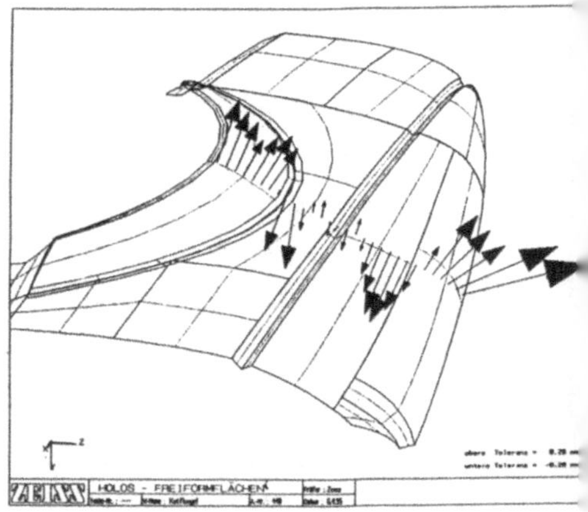


*Fig 15: Deviations represented as arrows.*

**4.3.3 Markings:** With this option, the deviations are displayed in colour tones. I.e. different deviation areas are marked with different colours. The advantage of this representation is that the deviation course can be seen very well due to the colours and when a large number of points is in question, a clear representation of the deviations is guaranteed. (Not represented graphically here due to the lack of coloured print.)

**4.3.4 Chromatic co-ordinates**: This representation, as is the markings option, is based on various colours, however in contrast to the markings, deviations are interpolated and displayed in colour in the area between the measured points as well. This makes it a particularly graphic type of representation. (Not represented graphically here due to the lack of coloured print.)

**4.3.5 2D Sections**: This option should be preferred if a line or a scanning line is measured in HOLOS. This line can then be displayed as a section, similar to a curve measuring program. I.e. the surface description is masked and only the nominal points and deviations are displayed in curve form (2D). The tolerance band, minimum and maximum deviation can also be displayed using a parameter input.

## 5. Mathematical Alignment and 3D Best Fit

Mathematical alignment is an absolute must for a part to be measured. It is necessary to define a common co-ordinate system between the measuring machine and the relative position of the workpiece on the machine.
With sculptured surfaces, this cannot normally be done with sufficient accuracy using conventional methods - regardless of their complexity - because the parts have no or only unfinished standard features. This means that the **form** and **position** errors are superimposed and will falsify information on dimensional integrity.

To minimise the influence of positional errors HOLOS incorporates an algorithm named **"3D Best Fit"** supplying the corrective values for optimising the co-ordinate system. The six degrees of freedom can be restricted at will in this best fit. Finally, to demonstrate the necessity of this algorithm, the graphical representation of the deviations of a measured area before and after its application are shows. (See Figs. 16 and 17).

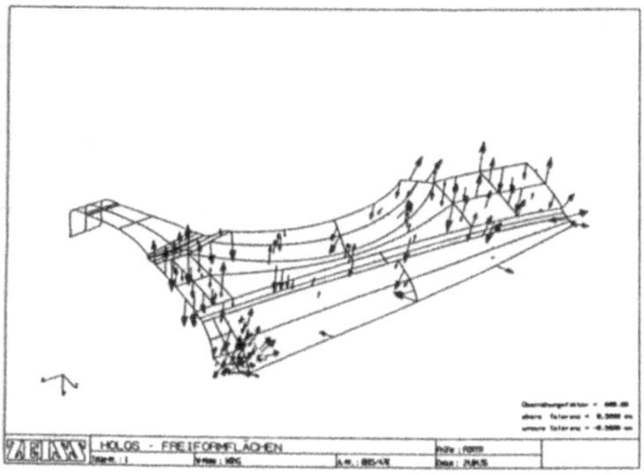

*Fig. 16: Deviations before the 3D Best Fit.*

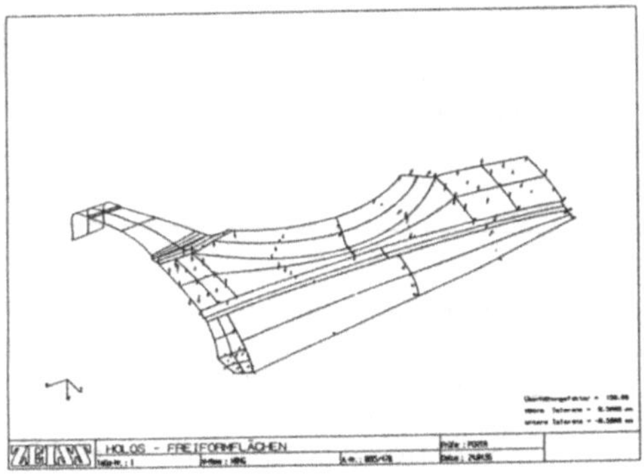

*Fig 17: Deviations after the 3D Best Fit.*

# 6. Conclusions

The purpose of this article was to introduce the reader to the subject of Sculptured Surface Metrology using Coordinate Measuring Machines in combination with the HOLOS software package. In order to clarify ideas, only the most important functionality has been shown.
With the introduction of CAD Technologies, the importance of CMMs has been increased enormously, making them an indispensable link in the chain of manufacturing technology. The development of new mathematical algorithms for the representation, manipulation and synthesis of curves and surfaces in combination with new measuring and other input devices, will enable the centralisation of the entire information related to a manufacturing process.

# 7. References


**[1] Neuman J. H.**
Coordinate Metrology. Technology and Application. (1990) Verlag Moderne Industrie.

**[2] W. Beyer; H. Kunzmann**
Today´s realization of traceability in dimensional metrology.
VDI Berichte 761, 1989. VDI Verlag.

**[3] R.Fröwis; C. Porta; U. Kleindienst**
Praktischer und theoretischer Nachweis der Richtigkeit von Freiformflächenmessunggen
VDI Berichte 1258, 1996. VDI Verlag.

[3] **C. Porta; R. Ohnheiser**
Conventional Procedures and Future Aspects for Measurement of Sculptured Surfaces with CAD-/CAQ-Methods.
VDI Berichte 761, 1989. VDI Verlag.

[4] **K.-H. Breyer ; H. J. Neumann ; C. Porta**
Holometrische Koordinatenmeßtechnik- Neue Ansätze zur umfassenden Messung beliebiger Formelemente.
VDI Berichte 1006, 1992. VDI-Verlag .

## Author

R. Fröwis; U. Kleindienst; C.G. Porta
Holometric Technologies
Softwareentwicklungs- und Consulting-Gesellschaft mbH
Technologiezentrum
Gartenstraße 133
D-73430 Aalen
Tel.: ++49 (07361) 560128 Fax.: ++49 (07361) 45187

# POMOS - POint based MOdelling System

*Karl-Heinz Häfele*

Research Centre Karlsruhe

**Abstract:** POMOS - POint based MOdelling System - developed in the Research Centre Karlsruhe is designed to close the gap between measuring devices and following CAX-applications. The capabilities of POMOS are focused on point operations, surface generation, surface analysis and interfaces from and to neutral file formats.

POMOS is not intended to replace a CAD-system but to be an add-on. Therefore typical CAD-functions such as fillets, intersections, projections etc. are not realised in the system.

POMOS is able to handle a large amount of digitised point data. This point data can be manipulated, modified, analysed and sorted. By defining three or more borders, the system is able to approximate free form surfaces. These generated surfaces can then be fitted together with either an edge or a smooth joint.

Before exporting the generated surfaces, the quality of these surfaces can be analysed using point-to-surface deviation, cutting lines, isophotes, reflection lines and curvature analysis.

## 1 Introduction

3D-digitising systems produce point clouds of varying sizes, orders and accuracy. These point clouds cannot usually be handled by CAD systems. In most cases it is more effective and less memory consuming to work with mathematically described surfaces (Monoms, Bézier, B-splines etc.). These surfaces can be handled, modified and manipulated more easily by CAD systems.

The POMOS system, developed in the Research Centre Karlsruhe, has been designed to generate free form surfaces on the basis of measured three dimensional point data.

In order to achieve an acceptable approximation of point clouds with surfaces in POMOS the process is divided into different steps. Figure 1.1 shows the basic steps.
As POMOS is independent from a specific measuring device, different point formats can be imported. Depending on the measuring device and on the measurement itself, the input data has to be processed. After defining the borders, curves or surfaces can be approximately calculated. As a single point can influence the surface quality, the generated surfaces have to be analysed. If the quality is not satisfactory, the bordering or the calculations have to be repeated with different parameters. Once the quality is at an acceptable standard, all the resulting data is available for exporting.

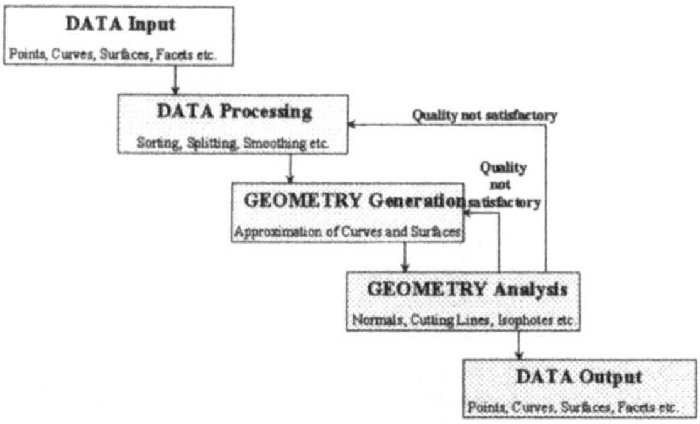


Figure 1.1: Processing steps

The development of POMOS is based on the following principles:

- Scattered point data has to be processed (automatically or manually)
- The coordinates of the measured point data should not be manipulated
- Generation of bicubic non rational polynomial surfaces
- Generation of non trimmed surfaces (except holes)
- Extended surface analysis methods
- Only very basic design functionalities (no typical CAD functions)

# 2 Main Features of POMOS

## 2.1 Data Input

POMOS is able to read in sorted and unsorted point data in various formats e.g. "ASCII X, Y ,Z" , VDAFS (points and point sets), NC, Stereolithography (polyeder or point cloud) and some binary formats.

Surfaces and curves which are already available, can be imported via VDAFS and used for further processing.

The IGES pre- and postprocessor is still under development and thus allows only specific entities to be imported and exported.

The adaptation of existing STEP pre- and postprocessor for POMOS is planned for the future.

The amount of data and the number of point clouds being imported is only limited by the main memory of the workstation.

## 2.2 Data Processing

The first step of data processing is to sort the point data. The importing of the point clouds automatically sorts the data into a grid using default values. As this sorting algorithm depends on the $2\frac{1}{2}$D property of the point cloud, the result may not always be acceptable. If the point data is a real 3D point cloud and totally unsorted, the point data has to be split manually into subareas with $2\frac{1}{2}$D properties. If the measuring device follows a measuring strategy, it is possible to recognise this and to produce scanlines and triangle networks even for a real 3D object [Hel95]. These different representations of a point cloud are shown in figure 2.1 [Hae96].

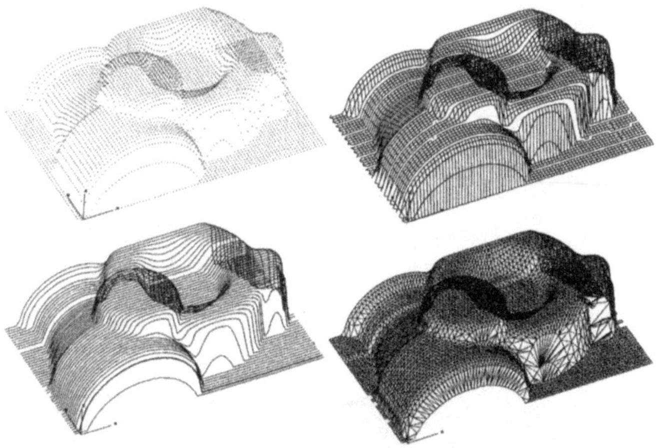

Figure 2.1: Different representations of point data (raw data, grid, scan lines, triangles)

All the point operations are applied regarding to the so called working box. The working box defines the current workspace and can be modified in size, position and orientation. Figure 2.2 shows operations applied on different workspaces. With the help of this working box all operations concerning point data can be limited to a certain area.
All different representations shown in figure 2.1 are based on point entities stored in the data base. If required, both the scanlines or the triangular mesh can be stored in the data base as either polylines or triangular meshes (polyeder). By defining the working box, the number and length of the scanlines converted to polylines can be controlled (see figure 2.2 lower left picture). The size of the triangle mesh is governed by the working box (see figure 2.2 lower right picture).

Some operations such as splitting, merging and removing of undesired areas are applied on unsorted point clouds. The detection of scanlines and the triangulation are also applied on raw data.

The intersections between point clouds and a plane can be calculated with both unsorted and grid data. The resulting intersection depends on the selected input data. An intersection with unsorted data requires certain tolerances and will produce a new intersection point cloud. An intersection applied on grid data will produce a polyline lying accurately within the defined intersection plane. Figure 2.3 shows two examples of intersection of grid data.

The first example shows an intersection where the plane is defined by drawing a directional line, which is then used to cut the grid data perpendicular to the screen. The second example shows multiple intersections parallel to the x-y plane. In this example, due to the complex

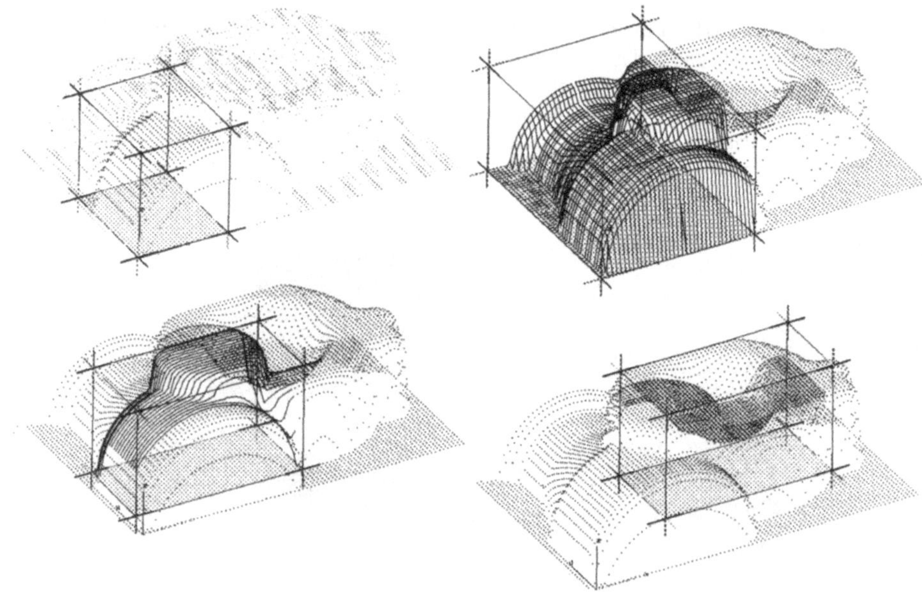

Figure 2.2: Operations applied on different working boxes (selection of raw data, gridding of point data, searching for scanlines, generation of triangles)

shape, some intersections have more than one intersection line. After converting to polyline entities such intersections will have more than one independent polylines as,a result.

Figure 2.3: Two examples of intersecting point clouds with planes

There are two functions available for working only on sorted data. If required the sorted data can be smoothed using either a Median or an Average filter algorithm.

More importantly and more difficult to implement are the methods to segment the point data. All of the segmentation methods implemented in POMOS are based on sorted data. There are different possibilities for segmentation of point clouds. One example implemented in POMOS is to interpret the angle between three points in different directions. The result of this process is a number of polylines offered to the user as possible edges. The first example in figure 2.4 shows the result.

Another way of segmentation is not primarily to look for edges, but to define regions. In the second example of figure 2.4 regions are found by interpretation of the reflection properties of the grid.

In both cases the result is strongly dependant on the quality of the measured data. Temporarily smoothing the data can sometimes improve the results.

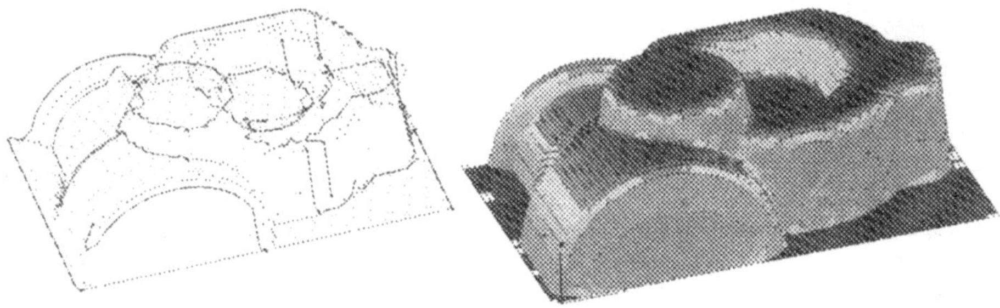

Figure 2.4: Segmentation of point clouds

## 2.3 Geometry Generation

The generation of geometry is based on polylines defined on a point cloud. The simplest way of creating polylines is to select two or more points on the grid. On one hand this polyline can be directly used to generate a curve or to border a surface. In this case only the selected points are taken into account. On the other hand the polyline can be projected onto the grid to obtain more points from the grid. The result of this operation will be a new polyline with more points. This polyline can also be used for generating a curve or bordering a surface. Figure 2.5 shows the projection of a manually defined polyline and the resulting curve.
Other possibilities to obtain polylines are intersecting planes with point clouds, converting scanlines to polylines and converting detected edge lines to polylines. In all cases the polylines can be edited and used for generating a curve or bordering a surface.

The easiest way to generate a surface based on a point cloud is shown in figure 2.6. The first step is to define roughly four polylines on the grid (figure 2.6 upper left corner). These polylines can be attributed by the number of segments and other optional attributes. This is the only interaction to create a surface. Internally the process is divided into different steps. If the suitable attribute is given the polylines will be projected onto the grid (figure 2.6 upper right corner). In the next step the borders are approximated to cubic polynomial curves (figure 2.6 lower left corner). In the last step the surface is calculated using the curves and internal points. The result is shown in the lower right corner of figure 2.6.

Depending on the number of polylines and the attributes, various surface constructions can be achieved. Figure 2.7 shows four examples. In the upper left corner an ordinary polynomial surface with four borders and two by two patches is calculated. By defining three borders and a peak point a triangle surface can be achieved (figure 2.7 upper right corner).

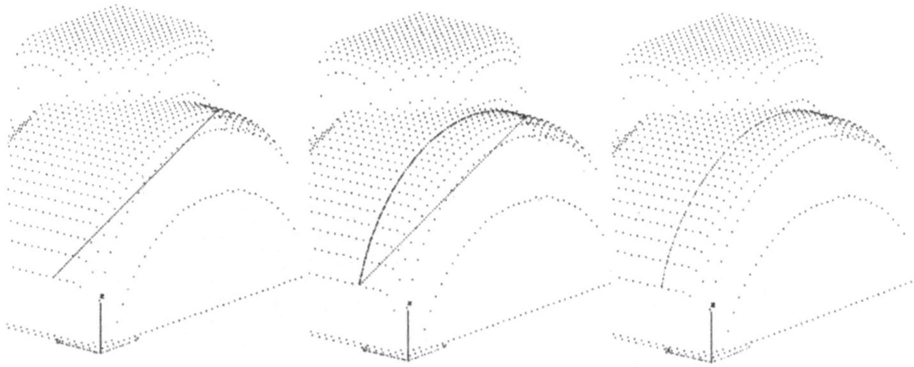

Figure 2.5: Defining a polyline and creating a curve (line by two points, projected polyline, curve)

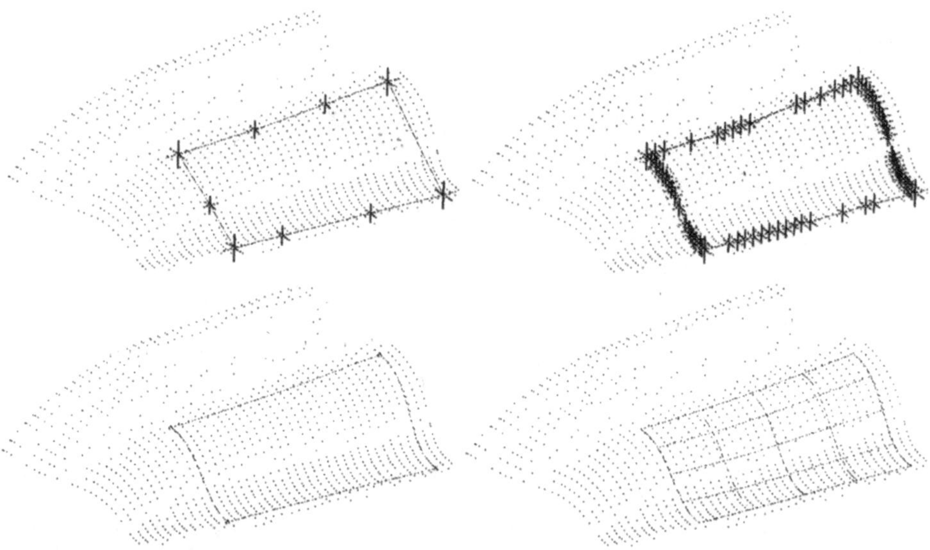

Figure 2.6: Defining four polylines and creating a surface (rough definition of four polylines, projecting the polylines onto the grid, calculating four curves, calculating the surface with internal points)

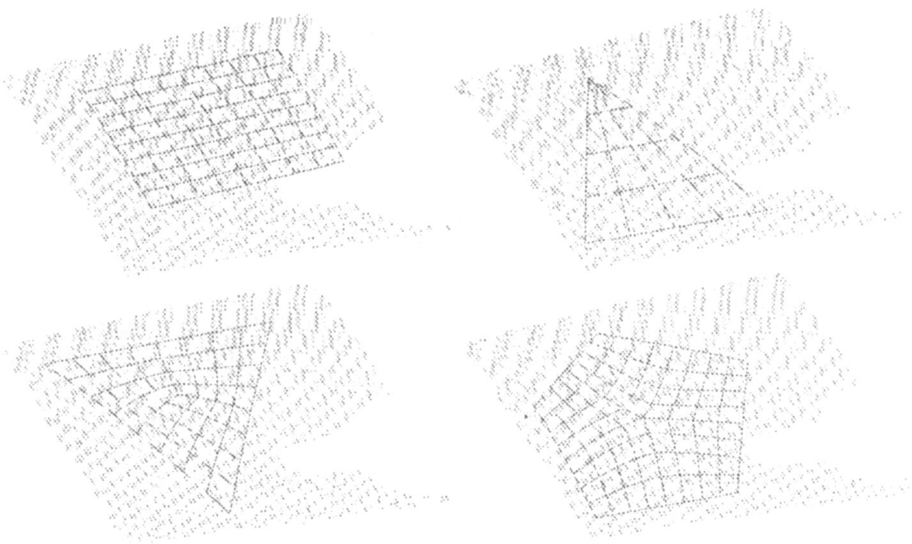

Figure 2.7: Examples of possible surface generations

Another possibility to create a triangular surface is shown in the lower left corner of figure 2.7. In this case only three borders are selected. The same algorithm can be applied for more than three borders [Hos92]. An example with five borders is shown in the lower right corner of figure2.7.

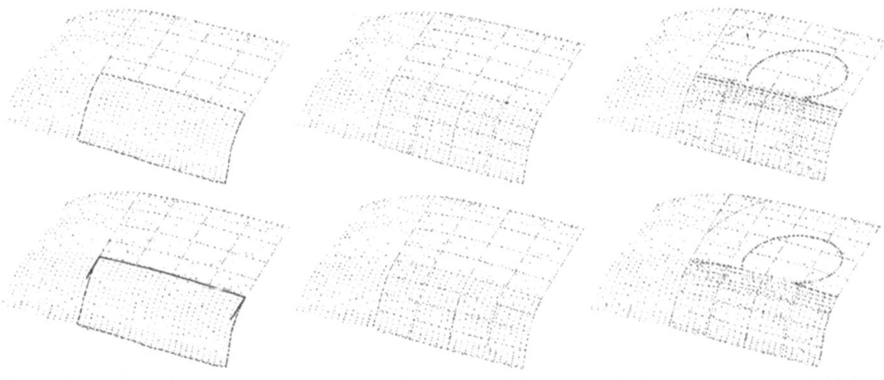

Figure 2.8: Connecting surfaces either with an edge (above) or with a smooth joint (below)

In order to complete a whole model it is necessary to connect surfaces. Figure 2.8 shows two possibilities to connect surfaces. In the upper row of figure 2.8 a surface is connected to an already approximated surface with $C^0$ continuity. The isophotes displayed in the upper outer right image jump at the surface border [Kum93]. The gap within this isophote present at the crossing of the border, indicates a ridge connection.

In the lower row, the same surfaces are connected with a $CG^1$ continuity. In this case the isophotes appear with an angle at the surface border. This indicates a smooth connection.

## 2.4 Geometry Analysis

As the quality of the generated geometry is strongly dependant on the quality of the measured point data, the geometry has to be analysed. POMOS offers a number of tools for analysing the quality of the surfaces and the joins between them [Hae95].

The first quality criteria is calculated with the surface generation: the maximum and average approximation error. These deviations between the points and the surfaces can be display in a colour map. The 'normal' distances between one point or the whole point cloud and the surface can be also calculated.

Especially in the edges and on the borders of the surfaces the normals can be an important quality criteria. The normals can provide information about dog's ears at free edges and borders as well as about the joins between surfaces..

Isolines (cutting lines) can be generated parallel to pre-defined planes. These are very useful for indicating irregularities on the surface.

In order to check the quality of the joins between surfaces, isophotes can be generated. Isophotes are lines of equal brightness created by global parallel lighting on the model. These lines allow to distinguish between $C^0$, $C^1$ and $C^2$ continuity.

A similar method to analyse the quality of surfaces with lighting is with the use of reflection lines. In this case the lighting source is not parallel, but simulates several linear lighting sources. Besides similar behaviour to isophotes, reflection lines display the smoothness of the surface more clearly, indicating any waves.

In order to analyse the curvature of surfaces, POMOS allows the calculation of gaussean and average curvature. Both can be displayed in either an iso-line or a colour map representation. The minimum and maximum radius of curvature is also available.

A simple method to analyse the behaviour of a surface are the Bézier control points. POMOS displays the control points as a control polygon. This can be visually interpreted.

## 2.5 Data Output

POMOS is able to export the geometry in the VDAFS Format. All entities of VDAFS version 1.0 are supported.

The point data can be converted into different file formats.

The IGES and STEP processors are still under development.

Polygons, which can be generated from surfaces and point clouds, are available in different virtual reality formats, in the DXF format and in the STL format.

## 2.6 Miscellaneous

The data base of POMOS is structured hierarchically and open for any kind of information. The structure of this data base can be graphically displayed and manipulated. Information about the data, such as data type, relationships, geometrical details etc. is available at any time.

Some simple functions to manipulate the geometry are available. Examples of these functions are: reverse, rename, copy, project, mirror etc.. Most of these functions work for points, polylines, curves, surfaces and polyeder.

In order to interact with the system POMOS offers all the possibilities to manipulate the screen layout, the colours, the language etc.. Pre-defined and user-defined views for each 3D-window are available.

Both the on-line manual and a more extended reference manual are available as hypertext documents.

# 3 The Examples

All the following examples have been supplied by external sources. Therefore the measurement devices and the measuring methods are unknown.

## 3.1 Example 1

The point cloud of example 1 consists of only 842 points. It seems that this point cloud was obtained by calculating points on a trimmed surface. The points calculated on the trim curve were at the end of the data file. Figure 3.1 shows the point cloud and the polylines, which border the surface. The size of the point cloud is approximately 110,5 units in the x-, 134 units in the y- and 46,5 units in the z-direction. The centre of the bounding box (local co-ordinate system) is located, regarding to the global co-ordinate system, at -11 units in the x-, -17,3 units in the y- and 4,5 units in the z-direction. The point cloud has not been manipulated.

After bordering the surface manually, POMOS calculated a non trimmed surface with five patches. The result is shown in figure 3.2 (left). In the middle of this multipatched surface a $G^1$ point was generated. At this point, the tangent plane of all patches are identical. At the borders of this multipatched surface, the patches are linked together with $C^1$ continuity.

In this case the maximum deviation between the point cloud and the calculated surface is very high (5,5 units). The reason for this is the low number of patches. In order to achieve a better approximation more patches are required. At the moment POMOS is not able to calculate the same surface structure with more patches. Future developments will solve this problem.

Other criteria of the surface quality are shown in figure 3.2 (right image) and figure 3.3. The Bézier control polygon shows the continuity effect of the $G^1$ point and of the $C^1$ borders. The cutting lines and the isophotes demonstrate the smoothness of the surface as well as the continuity effect at the borders.

## 3.2 Example 2

This example is larger than the previous with a point cloud of 4 101 points (figure 3.4 left). Through examination of the data it can be shown that the object is constructed of 49 scanlines

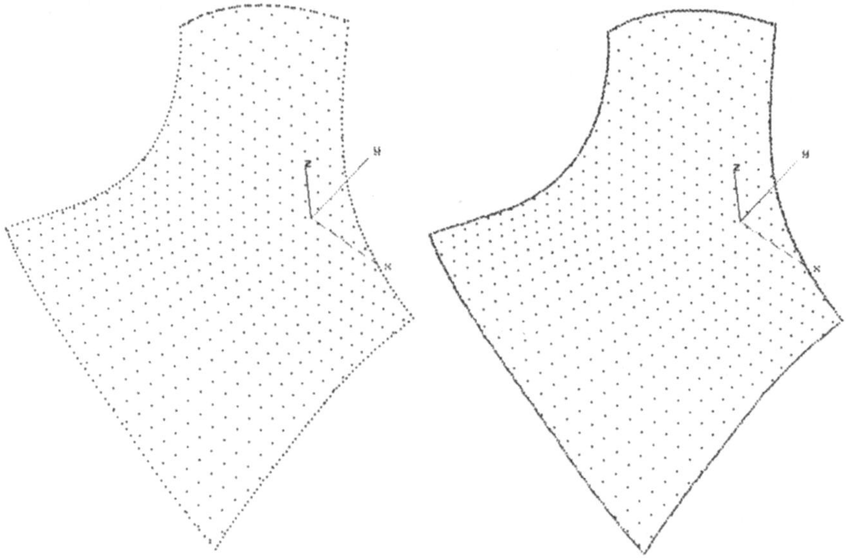

Figure 3.1: The point cloud of example 1 (left) and the manually defined borders (right)

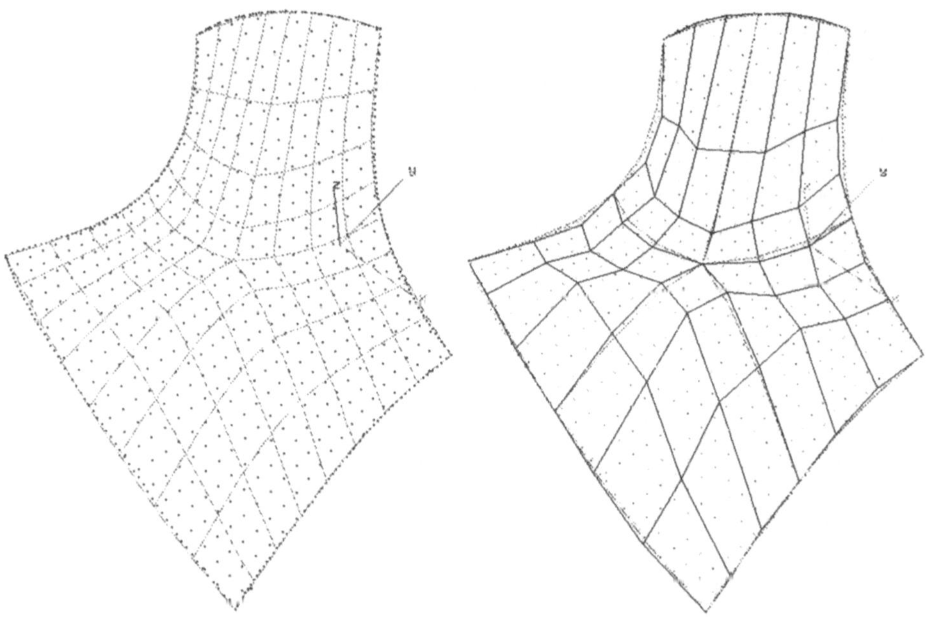

Figure 3.2: The generated surfaces (left) and the Bézier control polygon (right)

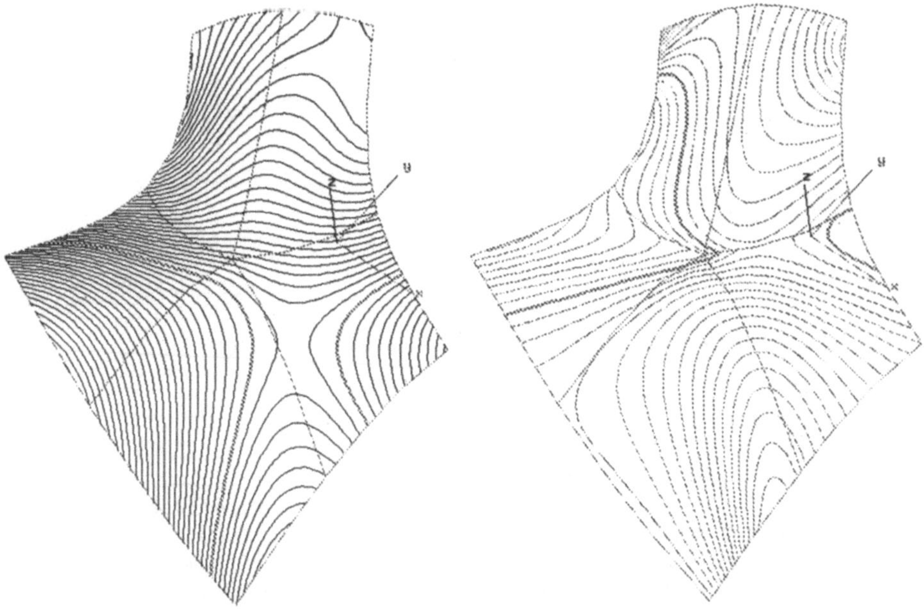

Figure 3.3: Cutting lines relative to the x-y plane (left) and the isophotes (right)

parallel to the x-z plane. These scanlines have been used by POMOS to generate a triangle mesh shown in figure 3.4 (right). This object contains 7 732 triangles. The dimensions of the point cloud are 81 units in x-, 50 units in y- and 26 units in z-direction.

Figure 3.4: Point cloud (left) and generated triangle mesh (right) of example 2

Attempts to segment this point cloud have been made with two algorithms shown in figure 3.4. The results are not satisfactory, but can help the user to generate border lines.
As large parts of this model are cylindrical, the optimum method of surface construction is to use 'surface of revolution'. At the moment this method uses a given curve and axis, which it uses to plot the surfaces. In order to use this method more accurately, the location of the curves and axes used in these calculations would have to be mathematically determined within the application. First approaches are shown in figure 3.5.

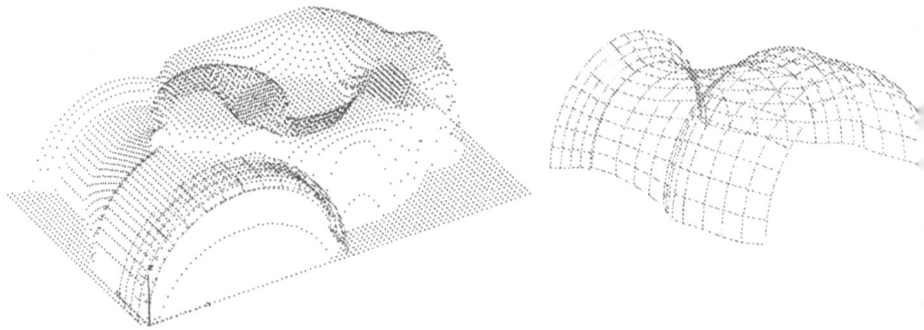

Figure 3.5: Surface of revolution generated by a approximated curve

## 3.3 Example 3

Example 3 has been measured with 40 scanlines relative to the y-z plane. The point cloud contains 1 827 points. This example shows the point density is dependant on the curvature of the object shape (figure 3.6 left image).The dimensions of this cloud are 124 units in the x-, 60 units in the y- and 355 units in the z-directions. The centre of the bounding box is found at the position of 2997 units in the x-, 481 units in the y- and 383 units in the z-direction. On the basis of this point data a triangular mesh with 3460 triangles can be obtained. This mesh is shown in figure 3.6 right image. In the generation of the surfaces the point data was not modified.

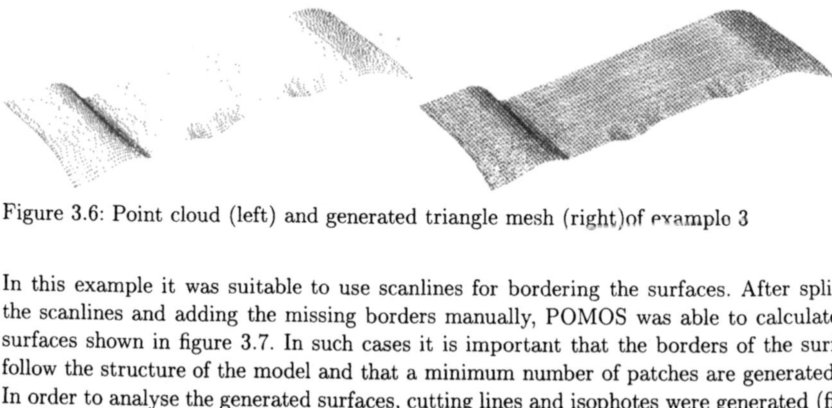

Figure 3.6: Point cloud (left) and generated triangle mesh (right)of example 3

In this example it was suitable to use scanlines for bordering the surfaces. After splitting the scanlines and adding the missing borders manually, POMOS was able to calculate the surfaces shown in figure 3.7. In such cases it is important that the borders of the surfaces follow the structure of the model and that a minimum number of patches are generated.
In order to analyse the generated surfaces, cutting lines and isophotes were generated (figure 3.8). Both the cutting lines and the isophotes do not indicate any irregularities. The isophotes show that all the patches and surfaces are linked together with a $CG^1$ continuity.
As can be seen in figures 3.7 and 3.8, a part of the surface has not been modelled (closest edge). This is due to greater deviations within the point cloud which form impressions on the object surface. The calculations of this area would mean an increased number of patches, which could cause problems with tangent continuity. In the near future POMOS will be able to solve this problem of connecting surfaces of differing numbers of patches.

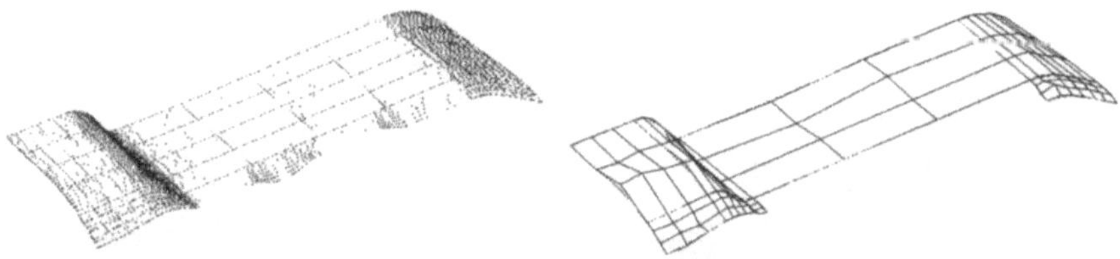

Figure 3.7: Generated surfaces in parameter representation (left) and the corresponding Bézier control polygon (right)

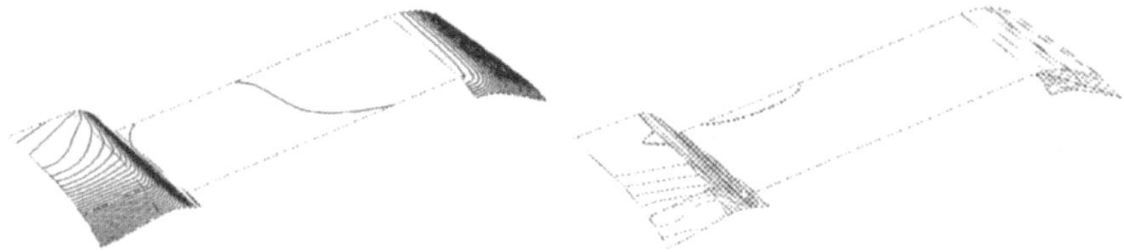

Figure 3.8: Cutting lines (left) and isophotes (right) calculated on the generated surfaces

## 3.4 Example 4

The point cloud of example 4 contains 95 986 points. The raw data and the sorted grid are shown in figure 3.9. The dimensions of the point cloud are 60,5 unit in the x-, 142 units in the y- and 162 units in the z-direction. The centre of the bounding box lies nearly at the global co-ordinate system (x -4,8 units, y 14,7 units, z -1,5 units)

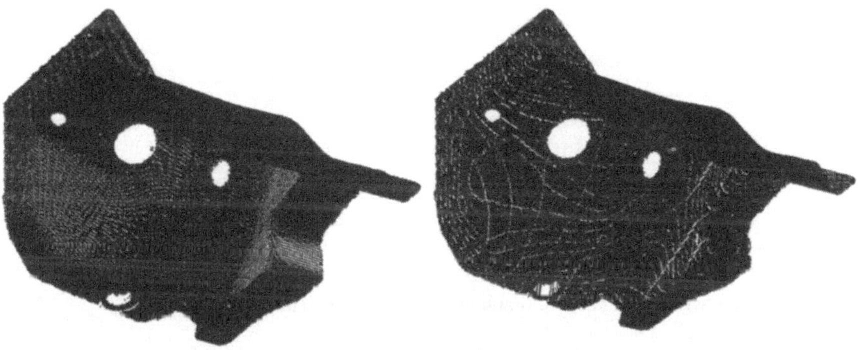

Figure 3.9: The raw data of example 4 (left) and the generated grid (right)

On analysis of the point cloud, 660 scanlines are found (figure 3.10 left). The scanlines are subdivided into four regions. Figure 3.10 left shows the bounding boxes of these regions. Figure 3.11 shows two of the scanned regions as triangular meshes in a shaded representation. The mesh on the left hand side of the figure was produced by a scan from the front of the object (relative to readers perspective), where the right hand mesh was produced by a scan from the rear. The blind spots of the rear scanned mesh can be clearly seen on the foremost surface.

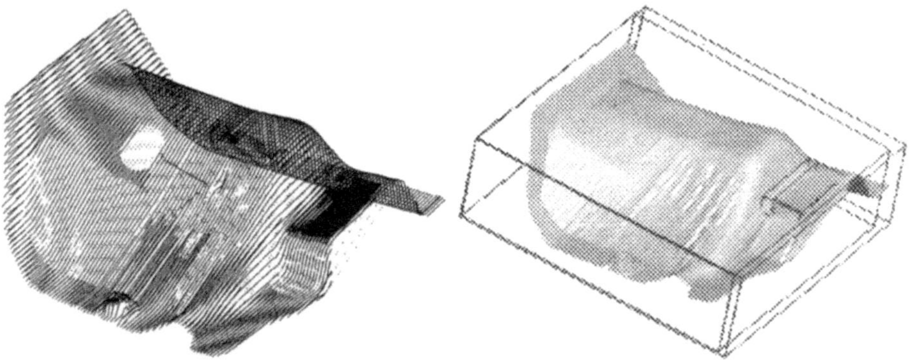

Figure 3.10: The found scanlines (left), displayed in a lower density and the scanned regions (right)

In order to manipulate and to generate geometry on the basis of triangular meshes, all submeshes have to be merged. A general function to merge arbitrary triangular meshes is under development.

Until this merging algorithm is working correctly, all manipulation and generation has to be done on the basis of the sorted data (grid) (see figure 3.9 right).

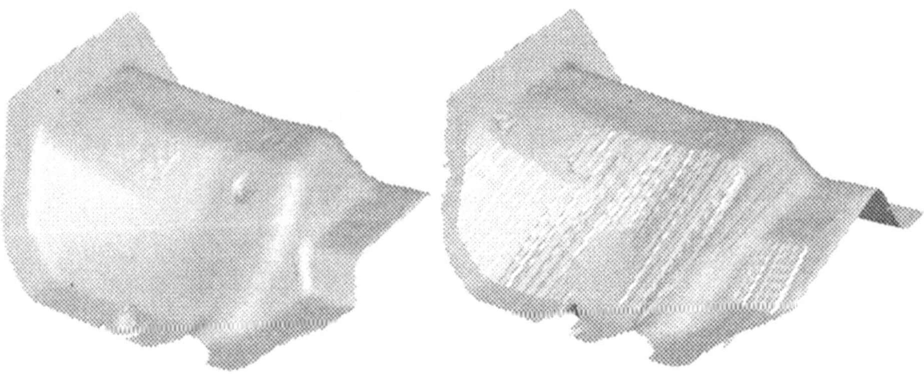

Figure 3.11: Two overlapping scans, displayed as a shaded triangular meshes

## 3.5 Example 5

This example consists of 105 709 points. The point cloud was measured using 307 scanlines. The point cloud and the scanlines are shown in figure 3.12 (both images do not have the whole point density).

After defining the outer and inner borders, the surface and the curves forming the hole can be calculated. Figure 3.13 shows the border lines as polylines and the surface with curves bordering the hole.

Figure 3.12: The point cloud of example 5 (left) and the found scanlines (right)

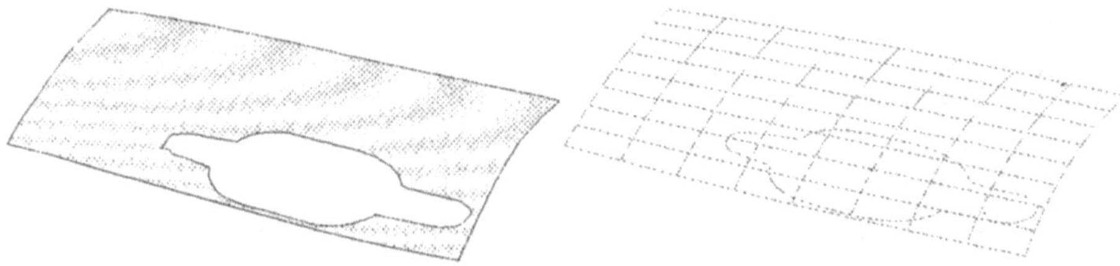

Figure 3.13: Polylines bordering the surface (left) and the calculated surface and curves (right)

The surface was calculated as a multipatched surface with four patches. The hole has been bordered by four curves. Two of them with two segments and two with four segments. The average deviation between the point cloud and the surface was 0.039 units. The maximum deviation was 0.861 units and was located at the border of the hole. Figure 3.14 shows greater deviation between the point cloud and the surface by darker colouring of the points and the smoothness of the surface by displaying isophotes.

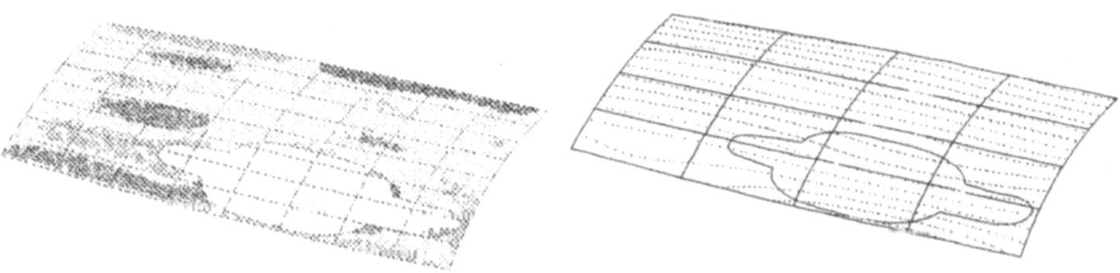

Figure 3.14: Deviation between point cloud and surface (left) and isophotes (right)

Another important criteria for the quality of the surface is the direction of the surface normals. Figure 3.15 (left) shows the normals. The number of normals depend on the wire density. The examination of the normals at the border areas is especially important to ensure the correct formation of the surface.

Both the surface and the curves were exported via VDAFS to the CAD/CAM system ANVIL 5000. Using this system the hole was generated by trimming the surface. The result is shown in figure 3.15 (right).

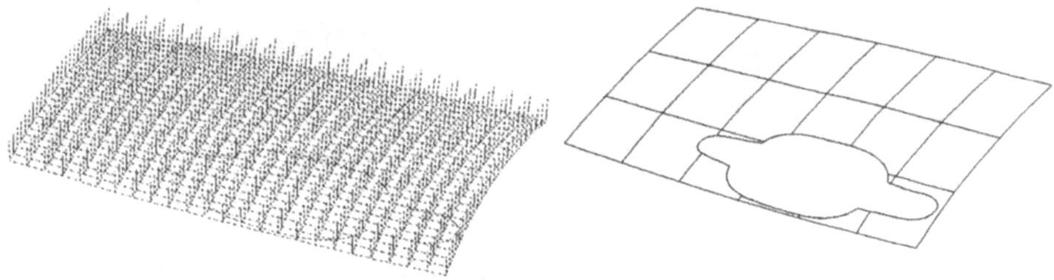

Figure 3.15: The normals generated on the surface (left) and the exported and trimmed surface in the CAD/CAM system ANVIL 5000 (right)

### 3.6 Example 6

Example 6 describes six cross-sections of a turbine blade relative to the x-z plane. Each cross-section has been measured with around 256 points. The dimensions of this point cloud are 25,6 units in the x-, 47,3 units in the y- and 29,4 units in the z direction. The raw data (1546 points) and the found scanlines (cross-sections) are shown in figure 3.16.

Figure 3.16: The raw data (left) and the detected scanlines (right) of example 6

The found scanlines can be used to generate curves. By splitting each scanline three times, four curves with four segments have been generated. As POMOS is not able to generate skinning surfaces, ruled surfaces between each cross-section have been calculated. Figure 3.17 (left) shows the approximated curves and figure 3.17 (right) shows the generated ruled surfaces. Of course ruled surfaces are not acceptable for modelling a turbine blade (isophotes in figure 3.18 left), but skinning algorithms are available in majority of CAD systems. So for this application POMOS was used to generate the skinning curves.

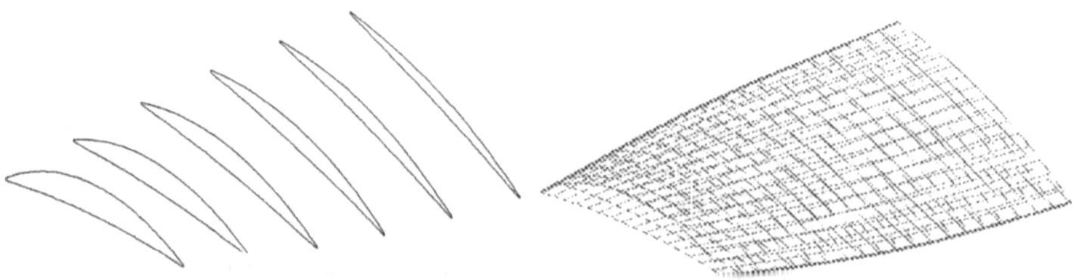

Figure 3.17: The generated curves (left) and the generated ruled surfaces (right)

For testing purposes the generated cross-section curves have been exported via VDAFS to the CAD/CAM system ANVIL 5000. With the help of this system a skinning surface based on the generated curves has been calculated. The resulting surface with isophotes is shown in figure 3.18 (right). Each cross-section curve has been subdivided into 11 segments to the fourth order. In the direction perpendicular to the cross-sections one segment to the sixth order has been generated.

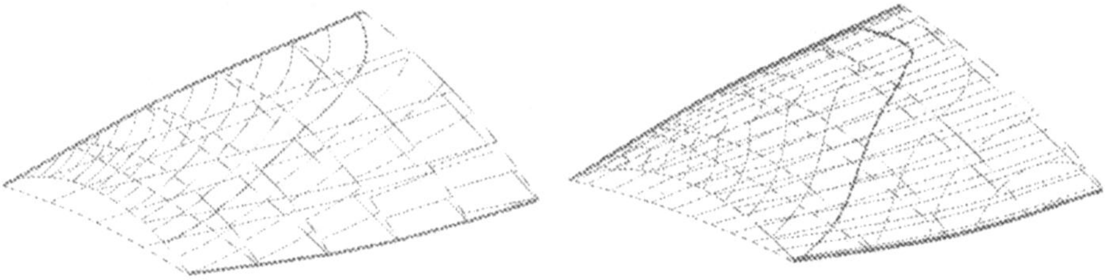

Figure 3.18: Isophotes generated on the ruled surfaces of POMOS (left) and on the skinning surface calculated by the CAD/CAM system ANVIL 5000 (right)

# 4 Conclusion

POMOS has been designed to generate curves and surfaces on the basis of measured 3D point data. Different measuring devices, measuring methods and various input data types forces this process to handle not only points, curves and surfaces, but a broader range of entities. The functionality of such a system should not be limited to simply the generation of surfaces, but also should be able to handle and manipulate point data and to examine the surface quality.

Each of the given examples has it's specific area of interest.

The point cloud in example 1 is formed with five boundaries. Without trimming, ordinary surface construction is limited to three or four borders. In order to prevent trimmed surfaces, a special surface construction (see figure 2.7 and [Hos92]) has been implemented in POMOS.

Example 2 requires 'surface of revolution' to complete the model. POMOS is able to calculate a 'surface of revolution' by given or approximated curves and a given axis. Opposing to this method and that of most CAD systems, it is desirable that both the rotating curve and the axis of revolution are calculated on the basis of the point data.

In order to minimise the number of surfaces and patches, example 3 requires to connect surfaces with differing segmentation of either $C^0$ or $GC^1$. Basically it is no problem to connect a surface with a higher number of patches to a surface with lower number patches (in the same parameter direction). The algorithms to complete this task are just in the testing stage and will be implemented in POMOS in the near future.

Example 4 has overlapped scan regions. As this point cloud has $2\frac{1}{2}$D property, the point data can be sorted. The triangulation of this point cloud will produce several triangular sub-meshes. To merge arbitrary triangular meshes is one of our aims. As these algorithms are not limited to $2\frac{1}{2}$D property, real 3D triangular meshes can be generated.

Example 5 shows a typical case for the trimming of surfaces. In this case POMOS is able to generate the surface and the trimming curves, but not to trim the surface. Exporting both the surface and the curves to a CAD system, the surface can be trimmed easily.

A similar situation is shown in example 6. In this case cross-sections were measured. POMOS is able to approximate the cross-sections by cubic curves, but not able to generate a skinning surface. As in example 5, the geometry was exported to a CAD system for the application of more specific CAD functions. Here the required skinning surface was calculated by the CAD/CAM system ANVIL 5000.

Working with point clouds is only one part of computer aided applications. It is not the aim of the POMOS system to compete with CAD packages. It is more vital to concentrate on the development of solutions to specific problems within the 'Reverse Engineering' process. With such a premise, systems such as POMOS should be used as an 'add-on' to or be integrated into an existing CAD package.

# References

[Hae95] K.-H. Häfele: POMOS - A System for Reverse Engineering, Rapid Prototyping in the Automotive Industries, 28th ISATA conference, September 1995, Stuttgart, Germany

[Hae96] K.-H. Häfele, M. Hellmann: 3D-Medatenaufbereitung und Weiterverarbeitung mit POMOS, 2. Workshop Optische 3D-Vermessung groer Objekte, Januar 1996, Esslingen, Germany

[Hel95] M. Hellmann: Abbildung von unsortierten Punkten des euklidischen Raumes auf ein Netz mit eindeutigen Nachbarschaftsbeziehungen, Diplomarbeit Fachhochschule Karlsruhe, September 1995, Karlsruhe, Germany

[Hos92] J. Hoschek, D. Lasser: Grundlagen der geometrischen Datenverarbeitung, B. G. Teubner, 1992, Stuttgart, Germany

[Kum93] H. Kummert; Erzeugung eines Freiformflächenmodells unter Berücksichtigung existierender Flächen, Studienarbeit, April 1993, Karlsruhe, Germany

# Author

Dipl.-Ing. K.-H. Häfele
Forschungszentrum Karlsruhe
Institut für Angewandte Informatik
Postfach 3640
76021 Karlsruhe
Tel.: ++49 7247 82 5745 Fax ++49 7247 82 5786
E-mail: haefele@iai.fzk.de

# Part II: Talks

# Some thoughts on Surface Reconstruction and shape control of planar curves, using target curvatures

*Alain Massabo*

MATRA DATAVISION

**Abstract:** Although it is a commonly accepted idea, Reverse Engineering does not mean Surface Reconstruction alone. Model controls using target criteria (Target Driven Design) can also be considered as belonging to Reverse Engineering. In the following presentation both aspects are discussed, the first one is illustrated by a general topic on the Surface Reconstruction current approach at MATRA DATAVISION, and the second one by an example of Target Driven Design applied to the control of planar curve shape by using a target curvature map.
***Key Words**: Surface Reconstruction, Target Driven Design.*

## 1 Introduction

Reverse Engineering is a general method for building or generating a model from its properties. Because, obviously and in general, there is **no biuniqueness** between model and properties, some **initialization** should be requested before **refining**. If we consider that measured points are "properties" of a CAD geometric model then Surface Reconstruction belongs to Reverse Engineering. I shall take the opportunity today of presenting another aspect of Reverse Engineering that, I think, will be more and more broadly and efficiently implemented in CAD/CAM/CAE systems, **model control via target properties**.
First let me give you an overview of how we treat Surface Reconstruction nowadays (it might change a bit in the future).

## 2 Surface Reconstruction at MATRA DATAVISION

The Surface Reconstruction process can be stated as:

*"given a set of points measured from a shape,*
*find the set of equations that "models" it".*

The main question is "*which subset of points corresponds to which equations?*" and its consequence is "*what are the "transitions" between equations?*". Transitions include the order of continuity between the different parts.
Another question is "*what is the measurement error?*". This leads to the choice between interpolation and approximation.
Our current answers to these questions is that it is the **user's responsibility with system assistance** i.e. the reconstruction is **not fully automatic.** The system assistance can be illustrated by characteristic curve extraction based on the chaining of points where a "$G^k$ discontinuity" is encountered.
Because the discontinuity is rarely located on measured points, these curves generally have to be considered as "suggestions", i.e. the user has tools to modify them and / or create new curves.

We make the hypothesis of high "density" scattered point set with a high cardinality (up to a few million of points). This characteristic needs special data structures and algorithms. Even if we have implemented such data structures, and despite hardware progress, I think this problem **still has to be explored** for even more points i.e. several millions.
In general the points are not really scattered but rather are organized in "curves". We therefore take advantage of the measurement process. A local grid organization of the points accelerates the process without being mandatory.

Our approach is not a revolution. As a matter of fact, because the process is not yet fully automatic, we think that the **complete set** of surface design functions has to be **accessible at all times.**
Hence, a rather **tight integration** in the CAD system of specific and (rather) efficient functions for reconstruction.
The set of measured points is treated as the object shape, some design functions are available with the set of points. Sets of (high density) points can be used to generate "curves on a support" as if it were a "continuous" surface. For instance, it the case of:

- parallel and perpendicular projections,
- "curve on a support" generations satisfying extremity constraints (curve blending).

As we try to automate as much as possible, we are still trying to improve the process of "subset" recognition and processing.

The reconstruction process follows the classical computer aided surface design approach:

- "topology" definition that gives sets of characteristic curves, subsets of points and continuity definition,
- generation of curves and boundary cleaning that gives nets of compatible curves,
- filling by and fitting on sets of patches,
- optimization of intermediate results.

The topology definition is based on a "**declarative**" approach of characteristic curves, areas - each of them corresponding to one treatment ("subset") - and the order of continuity between areas. The system assistance proposes some characteristic curves that can be used by the user as suggestions for surface boundaries (see above). The result is often a set of curves that has to be "cleaned" (trimmed and / or modified to match at vertices) to constitute nets of compatible curves.
These steps are sensitive since they influence the quality of the first result. Fortunately, further steps may be used to make necessary corrections.
The fitting and filling steps - let us call them *skinning* -produce a set of patches that can be further improved (optimized) by using standard CAD functions .
Optimization, if need be, keeping continuity order, is:

- either quasi-automatic (body healing),
- or manual as any standard design process such as surface extension and trimming (intersection), blending with $G^2$ continuity.

Adapted visual tools for deviation checks are based on a color map of distances from measured points to resulting surfaces. We have tuned the distance algorithm to a high number of points.

As a (partial) conclusion, the tight integration of:

- specific functions,
- adapted ergonomy - millions of points, visual and real time checks,
- advanced CAD functions for surface design,

constitutes a **"nice" Surface Reconstruction tool.**

## 3 Target Driven Design

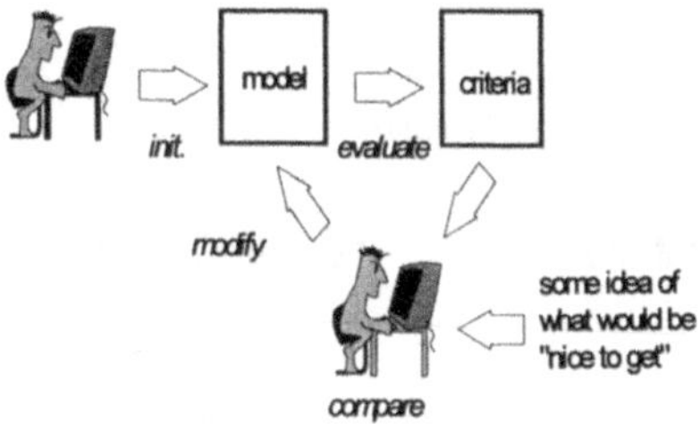


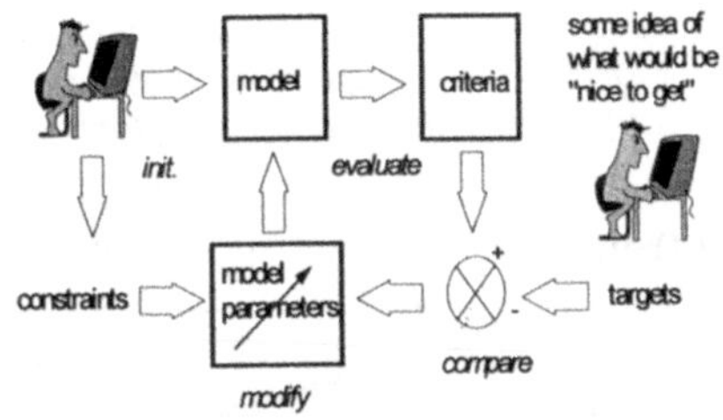


fig. 1-model control using direct approach

fig. 2-model control using "reverse" approach

Figures 1 and 2 illustrate two different methods for designing. They are not exclusive. In the direct approach users [try to] reach the goal by trial and error on the data model. There is often a lot of parameters, i.e. degrees of freedom, hence a lot of "manual loops" especially in Styling.
In the "reverse" approach after having been given by users an initializing model, target properties and constraints the system refines the model until a criterion is reached The loops are automatic and should the process converge, the users will get what they want.

In the example on the right the user builds the three circles tangent to two sides of the triangle and inside it.
The target is to find three radii that makes all three circles tangent to one another. The constraints are to keep them inside the triangle.
This can be expressed by an algebraic criterion.

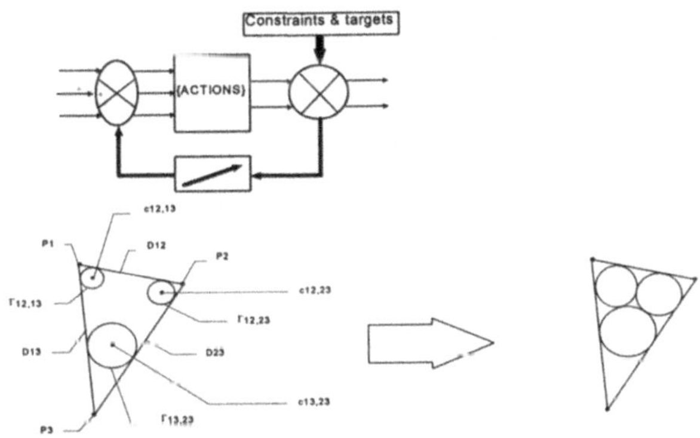


We call a design based on the "reverse" approach a **target driven design**.

Considering planar[1] piece-wise generated curves, we have experienced a way to improve their aesthetic quality by trying to solve the following "problem":

> *"from a given (piece-wise generated) curve and boundary constraints, try to reach the look of a given target curvature map by refinement "*

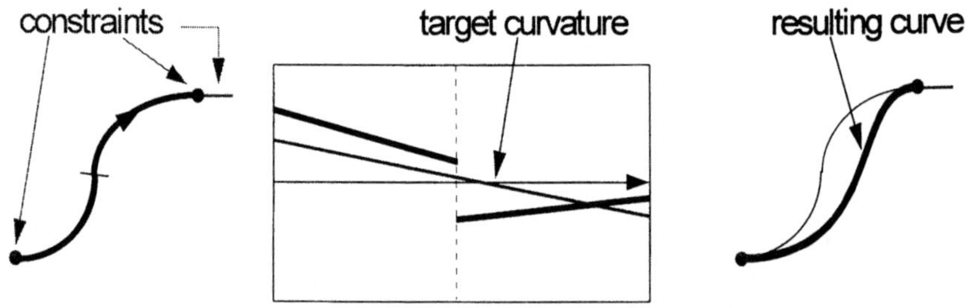


The **look** is more **important** than the **value** except at the extremities where sometimes the value is necessary for continuity reasons.

This method may not be the most elegant one but it is interesting for learning ...
The problem is:

- non linear,
- non convex,
- sometimes "over constrained",

and still under study even if it is "operational". This function is available in our system.

The following definitions are used:

$\{curves\} = C(t)$ and $s$ its arc length, where $t \in [0,1]$, $u = \dfrac{s(t)}{L(C)} \in [0,1]$ and $L = length(C)$

The $curvature\ map = \rho(t)$ or $\rho(u)$ and the $target\ curvature\ map = \bar{\rho}(t)$ or $\bar{\rho}(u)$.

The question is "*how can we define the look ?*"
In other words "*which criterion has to be satisfied ?*"

---

[1] The curves are planar to avoid any torsion influence

We have studied the minimization of the three following criteria:

$$\Pi(C)=\int_0^1 \left(\rho(t)-\bar{\rho}(t)\right)^2 dt$$ parametric criterion, not "meaningful".

$$\Gamma(C)=\int_0^1 \left(\rho(t(u))-\bar{\rho}(u)\right)^2 du$$ geometrical criterion, "expensive".

$$\Omega_\alpha(C)=\frac{\Pi(C)+\alpha\cdot\Theta(C)}{1+\alpha} \quad ;\ 0\leq\alpha$$ "compromise", that we have chosen.

with $$\Theta(C)=\int_0^1 \left(\frac{ds}{dt}-L(C)\right)^2 dt$$ parametric criterion, "helpful".

$\Pi(C)$ is **not meaningful** because it is **not** an **intrinsic** criterion. It looks like the "parameter distribution" between the two curvature maps "absorbs the criterion".

$\Gamma(C)$ is meaningful, but **several** *s(t)* can achieve it. This criterion is **not "really convex"**, convergence is difficult to achieve. We must add a stability factor, i.e. to "freeze" *s(t),* hence $\Theta(C)$.

$\Theta(C)$ is helpful because it "**improves**" the **convexity** of the space. As a matter of fact, it should be a constraint $\Theta(C) = 0$.
It is in a certain way "**orthogonal**" to $\Gamma(C)$, but not to $\Pi(C)$. This may cause problem because of an occasional linear dependency between $\nabla(\Theta(C))$ and $\nabla(\Pi(C))$.

Nevertheless, we have chosen $\Omega_\alpha(C)$ because of easier numerical computation[2]. Hence, we try to solve:

$$\underset{C}{MIN}\left(\Omega_\alpha(C)\right) \text{ with } \Pi(C)\approx 0 \text{ and } \Theta(C)\approx 0$$

and the constraints at each extremity of $C$ can be $G^k$ , k = 0, 1, 2

A Newton-like method is used, some heuristics are also implemented such as for example *a* which is tuned to "help" the iteration process, i.e. its value maximizes the minimal eigenvalue of:

$$\nabla^2\Omega_\alpha(C)$$

The best result is reached when $\Pi(C)$ and $\Theta(C)$ are equal to zero or almost zero ! But sometimes problems occur, such as the following:

$$\nabla\Omega_\alpha(C)=\nabla\Pi(C)+\alpha\cdot\nabla\Theta(C)=0 \quad \text{with} \quad 0\ll\left\|\nabla\Pi(C)\right\|$$

[2] Note that we could have taken $\Omega_a(C)= (1-\alpha)\cdot\Pi(C)+\alpha\cdot\Theta(C)$ where $\alpha$ belongs to [0,1].

It may happen that boundary constraints are **incompatible** with the existence of a solution, i.e. the domain is empty.

When it is not possible to converge we try to solve:

$$\underset{C}{MIN}\left(\mathrm{M}(C)\right) \text{ where } \mathrm{M}(C) \text{ is a majorant of } \Gamma(C)$$

$$\text{with } \left\| \frac{\Gamma(C) - \Pi(C)}{\Gamma(C)} \right\| \leq \delta \text{ (given value)}$$

$\Theta(C)$ has to be small to avoid cusps ($ds/dt = 0$) and also for computational reasons.

It is hoped that the majorant will decrease to a value less than was reached by $\Omega_\alpha(C)$, and by definition so will $\Omega$.
It is **only** a **heuristic** to find a better solution in case of problem, i.e. to search for a better local minimum. I am aware that it can be "discussed" ... !

Remarks

- When a solution exists, it is independent of $\alpha$, and even of the initial $C$ !!
- We have not yet implemented the control of the distance between the initial curve and the final one.
- Since $\rho(s)$ gives $C$, it may exist a better way to define a "look" ...
- "Dual problems" (?) of the direct approach are encountered:
  - target and "distance" (criteria) definitions,
  - constraint definitions,
  - target control,
  - existence of solutions.

The control of the distances between initial and final curves could be achieved by inequality constraints. The solution would, in general, be located at the boundary of the domain.

"Dual problems" must not be neglected.

- targets and distances (i.e. the criteria) have to be carefully defined,
- control of the targets has to be easy for the user.

The reverse approach is interesting if it **reduces** the number of degrees of freedom. The targets "replacing" the parameters of the direct approach, their definitions should not lead to more "manual loops". If the problem is over-constrained, then we can minimize a global criterion based on a balanced function of targets and constraints, i.e. we accept a compromise between targets and constraints. If the problem is under-constrained, optimization criteria are set in addition to the targets. Therefore there are several criteria, which have also to be "balanced" (weighted). In both cases the user has to keep an easy control for balancing.

## 4 Conclusion

The main issues of Surface Reconstruction are to find subsets of points which correspond to single $C^{\infty}$-continuous generations with the continuity orders between them and to approximate each of them with piece-wise sufficient continuous sets of equations. Any discontinuity that is approximated by a span generates artifacts that decrease the quality of the results and higher is the order of discontinuity less is its impact.
The main issues of target driven design are to find meaningful criteria and to reduce the number of parameters for allowing users to keep the control.
Reverse Engineering is not only Surface Reconstruction.
Reverse Engineering as model control using target properties
- seems to be a promising technique,
- but there are still some pending problems.

In both methods there is a room for further work ...I can only say we are still at the beginning and a lot of work remains to be done!

## Author

Dr. Alain C. Massabo
MATRA DATAVISION GmbH
53, avenue de l'Europe
13082 Aix-en-Provence CEDEX 2
Tel.: ++33 42 52 20 00 Fax.: ++33 42 52 21 99

# Surface design based on desired shadings, isophotes or curvatures

*Roger Andersson*

Mid Sweden University

**Abstract:** Methods for surface design based on certain desired properties will be presented. Working in this way enable CAD-users to entirely concentrate on prescribing properties, leaving the task to fulfil them to the computer. In this way, the operator may almost instantaneously attain design solutions than would takes days or weeks with traditional means - if reached at all.

A user may, however, prescribe properties for which there does not exist any physically realisable surface having them, as will be illustrated. From a mathematical point, this is to be expected, since most styling properties gives rise to nonlinear partial differential equations that in general, does not have smooth solutions. It is shown that for several visual- and curvature properties, there are unique perturbation solutions admitting a user, adhering to a few basic principles, to work rather freely.

## 1 Introduction

The purpose of this paper is to present some cases of surface design based on desired properties. In this recent method, a user specifies properties he wants see fulfilled, but leaves the actual modification of the surface to attain a prescribed property to the computer. More precisely, the wanted relationships are interpreted as equations that are solved numerically on a computer. Just as other parts of the design process have attained solutions much closer to the optimum by benefiting of often massive FEM-computations, similar or even greater impact is to be expected within areas of surface design where shape is important.

The present paper focuses on certain visual properties like a desired distribution of brightness intensity, isophotes or curvatures that are common means in the analysis of surface quality. For the precise definitions of these concepts and their use, we refer to [5]. The scope of the method is not confined to these cases. In principle, it would enable a user just to specify properties, while the fulfilment of them are accomplished by suitable software systems. Here, "just to specify" does not mean that we consider this a simple task, quite on the contrary! From our experience mainly earned within the automotive industry, we have learned that to obtain good styling results is a challenging task that requires considerable talent and artistic feeling. The importance of the method is it releases the user from an extremely time-consuming and tedious process that often ends due to fatigue rather than of being close to an optimal solution.

Most of the results in the paper concerns conditions ensuring existence of a useful solution to the problems considered. In many other application-oriented computations, existence of the object looked for may surely be taken for granted. In surface design, however, it is too easy to prescribe properties for which there is no physically realizable surface having them!

If these cases are not recognized, one may easily waste lots of computer time just to produce garbage. Even after such a failure has occurred, and one may trace back all steps, it is not easy to find out why the failure did occur.

## 2 The nature of the problems

In general, visual properties of surfaces depend on the behaviour of their surface normals, which in turn depend in a non linear way on first order partials. To obtain a surface with desired visual properties then, one has to solve a first order, nonlinear partial differential equation. Since $C^1-$ or $C^2-$surfaces are generally required, we must exercise some care in posing the problems, not to rule out the possibility to get smooth solutions to the equations right at the outset. In particular, one has to determine what parts of the boundary that one is allowed to fix and what parts that must be free to move. It is scarcely possible to make this decision by intuition only. Rather the system must assist the user in this regard.

Since curvatures depend on partials up to and including the second order, to find surfaces with a desired curvature, one is led to consider partial differential equations of the second order. As in the case of visual properties, these equations are non linear. The type of these equations depend on the kind of curvature. For mean- and Gauss curvature, the resulting equations are elliptic. As a result, for these curvatures, we may fix all of the boundary, and in doing so, the solution, if it exists, will be uniquely determined. Roughly speaking, problems of this kind are fairly robust and allow smooth solutions under reasonably general conditions.

For normal curvatures and curvatures of non normal sections, the corresponding differential equations will be parabolic in the broad sense. Problems of this kind are much more fragile, and in this generality encompass both problems for which one may find smooth solutions as well as ones, which in general do not admit a smooth solution.

We will present conditions ensuring smooth solvability. Under these conditions, we will get equations, parabolic in the narrow sense, similar to the well known heat equation. The conditions are rather restrictive, but allow us to treat important types of surfaces like, in the automotive industry, hoods, roofs and trunk-lids to mention a few. Even so, however, due to the evolution character of the equation, the problem is asymmetric with regard on smoothness: smoothness along integral curves to a certain vector field is higher than cross them.

When these conditions are violated, one will instead get equations that change type between backward- and forward parabolic. Equations of these types are ill-famed in mathematics since at least the beginning of the century. Despite from the question whether there are smooth solutions at all, which is still unsolved, this is an ill-posed problem which makes the numerical treatment hard: even *very* small changes in the right-hand side may give rise to *big* changes in the solution.

## 3 Image intensities and isophotes

Here we want to modify a surface $S_0$, having image intensity $I_0$ into new surface $S$, having image intensity $I$. Not to hide the essentials behind complicated formulas, we will assume that $S_0$ and $S$ are function surfaces. Thus $S_0 = \{(x, y, u^{(0)}(x, y)) \in \mathbf{R}^3 \mid (x, y) \in \overline{\Omega}\}$ and its surface normal $\overline{N}_0(x, y)$ at the point $P = (x, y, f(x, y))$ is given by

$$\overline{N}_0(x,y) = \frac{(-u_x^{(0)}(x,y), -u_y^{(0)}(x,y), 1)}{\sqrt{1 + (u_x^{(0)}(x,y))^2 + (u_y^{(0)}(x,y))^2}}$$

According to the diffuse reflection model, see e.g. [4], the image intensity $I_0(x, y)$ at the point $P$ caused by a single, distant light source $\overline{L}$ whose direction is given by a unit vector $\mathbf{l}$ is proportional to the cosine of the angle between $\overline{N}_0(x, y)$ and $\mathbf{l}$, see [4], p 723-724. Since we will not consider light sources with directions parallel to the xy-plane, we may express the Unit vector $\mathbf{l}$ as

$$\mathbf{l} = \frac{(a, b, 1)}{\sqrt{1 + a^2 + b^2}}$$

and in these notations, the image intensity appears as

$$I_0(x,y) = \frac{1 - a\,u_x^{(0)}(x,y) - b\,u_y^{(0)}(x,y)}{\sqrt{1 + a^2 + b^2}\sqrt{1 + (u_x^{(0)}(x,y))^2 + (u_y^{(0)}(x,y))^2}}$$

This is the expression one have to evaluate in several points $(x, y) \in \Omega$ in order, e.g. to create a shaded picture of the surface. It is also the point of departure to produce *isophotes*, which are curves of equal brightness, see [5], p 35. Families of such curves are frequently used in current surface analysis.

In the method we are discussing, one turns things round. Thus one modifies the intensity function $I_0$, either directly or indirectly by modifying some family of isophotes, until it looks as we like, giving rise to a new desired intensity $I$. Then we are looking for a smooth surface having this intensity. Usually, one also wants to keep as much as possible of the boundary fixed. So we want to find a smooth function $u$ in $\overline{\Omega}$ such that

$$\frac{1 - a\,u_x(x,y) - b\,u_y(x,y)}{\sqrt{1 + a^2 + b^2}\sqrt{1 + (u_x(x,y))^2 + (u_y(x,y))^2}} = I(x,y) \qquad \text{in } \Omega$$

$$u(x,y) = u_0(x,y) \qquad \text{on } \Sigma$$

where $\Sigma$ is some part of the boundary of $\Omega$.

This is a boundary value problem for a first order, non linear partial differential equation, known as the *image irradiance equation*. As a result, our surface modification problem has a solution if and only if this boundary value problem has. Unfortunately, this is generally not the case! To cite [6], smooth solutions ... are in fact "isolated accidents" for general equations of this kind. Indeed, Fig. 3 - Fig. 5 furnish examples in which the equation lacks a smooth solutions, whereas Fig. 1 - Fig. 2 on the other hand show a case admitting one.

To create reasonably reliable software for this property-driven design, we need to know when a useful solution may be expected and also help the user to put his requirements in a way that will not rule out the possibility to find such a solution. These matters are considered in some detail in [3]. Here, we will list some of the most important conditions ensuring an useful solution, referring to [3] for their precise formulations, the complete set of conditions and for proofs.

Concerning the boundary conditions, the following vector field plays a main role. With $u^{(0)}$ our original function, form

$$A = \frac{a(1+(u_y^{(0)})^2) - u_x^{(0)}(b\,u_y^{(0)}-1)}{\sqrt{1+a^2+b^2}\,[1+(u_x^{(0)})^2+(u_y^{(0)})^2]^{3/2}}$$

$$B = \frac{b(1+(u_x^{(0)})^2) - u_y^{(0)}(b\,u_x^{(0)}-1)}{\sqrt{1+a^2+b^2}\,[1+(u_x^{(0)})^2+(u_y^{(0)})^2]^{3/2}}$$

and let $\zeta = (A, B)$ be the plane vector field with components $A$ and $B$ . Furthermore, let $\nu$ be the exterior normal to the boundary of $\Omega$ and let $\Gamma_+$ and $\Gamma_-$ consist of the boundary points in which $\nu$ exists and for which the angle between $\nu$ and $\zeta$ lies between 0 and $\pi/2$ and between $\pi/2$ and $\pi$ respectively.

The relevant measure of size of the difference between two image intensities $I$ and $I_0$ involves not only the functions themselves, but also their derivatives up to and including the second order. For this reason, but also since it will be needed at other places in this paper as well, we introduce the space $C^k(\overline{\Omega})$. It consists of all continuous functions $v : \overline{\Omega} \to \mathbf{R}$ that has continuous partial derivatives of order $\leq k$ in $\Omega$, where all these partial derivatives can be extended into continuous functions on the closure $\overline{\Omega}$ of $\Omega$. Here, $k$ is a non-negative integer and $C^0(\overline{\Omega})$ means just the continuous functions on $\overline{\Omega}$.

The size, denoted $\| v \|_{C^k(\overline{\Omega})}$ of a function $v \in C^k(\overline{\Omega})$, is defined by

$$\| v \|_{C^k(\overline{\Omega})} = \sum_{j=0}^{k} \max_{p\geq 0, q\geq 0, p+q=k} \sup_{\Omega} \left| \frac{\partial^{(p+q)}}{\partial x^p \partial y^q} v \right|$$

for non-negative integers $p$ and $q$. Thus to compute e.g. $\| v \|_{C^2(\overline{\Omega})}$ we must compute all partial derivatives of the first- and second order $\frac{\partial v}{\partial x}$, $\frac{\partial v}{\partial y}$ and $\frac{\partial^2 v}{\partial x^2}$, $\frac{\partial^2 v}{\partial x \partial y}$, $\frac{\partial^2 v}{\partial y^2}$, find the numbers

$$p_1 = \sup_{\Omega} \left| \frac{\partial v}{\partial x} \right|, \; q_1 = \sup_{\Omega} \left| \frac{\partial v}{\partial y} \right|$$

$$r_2 = \sup_{\Omega} \left| \frac{\partial^2 v}{\partial x^2} \right|, \; s_2 = \sup_{\Omega} \left| \frac{\partial^2 v}{\partial x \partial y} \right|, \; t_2 = \sup_{\Omega} \left| \frac{\partial^2 v}{\partial y^2} \right|$$

as well as $m_0 = \sup_{\Omega} | v |$, $m_1 = \max(p_1, q_1)$ and $m_2 = \max(r_2, s_2, t_2)$, and then get $\| v \|_{C^k(\overline{\Omega})}$ by adding $m_0$, $m_1$, $m_2$ together.

With these notations behind us, we may now present the conditions:

1. Choose $\Omega$ in such a way that $\zeta$ is not a tangent to the boundary of $\Omega$
2. Fix $u$ along $\overline{\Gamma}_-$ *or* along $\overline{\Gamma}_+$ and let $u$ be free along the rest of the boundary
3. Choose a light source direction in such a way that in $\overline{\Omega}$

$$a\,u_x^{(0)} + b\,u_y^{(0)} < 1$$

$$(u_x^{(0)}, u_y^{(0)}) \neq (-a, -b)$$

4. Keep $\| I - I_0 \|_{C^2(\overline{\Omega})}$ small enough

The first part of the third condition means that no part of the surface we are going to modify is in shadow and the second part means that at no point of $\overline{\Omega}$, the surface normal be parallel to the light source direction.

We will now present two examples, illustrating these conditions and above all, the influence of the last one. In both examples, the function $u_0$, defining the original surface $S_0$, is given by

$$u_0(x,y) = \sqrt{1 - (\frac{x}{4})^2 - (\frac{y}{2})^2}$$

and the light source direction is determined by $a = b = 1$, so the unit vector pointing towards the light source is $\mathbf{l} = \frac{1}{\sqrt{3}}(1,1,1)$. For $(x,y) = (\frac{16}{\sqrt{21}}, \frac{4}{\sqrt{21}}) \approx (0.87, 3.49)$, the surface normal is parallel to the light source direction, but in choosing $\Omega = [0, 2.2] \times [0, 1.6]$, we avoid this point.

In both cases, we form the new intensity $I$ from that of the original surface $I_0$ by $I = I_0 + c \cdot J$, where

$$J(x,y) = \frac{(x \cdot (2.2 - x) \cdot y \cdot (1.6 - y))^2}{(1.1 \cdot 0.8)^4}$$

This function vanishes along the boundary $\partial\Omega$ of $\Omega$, i.e. along the edges of the rectangle, and has a maximum equals 1, at the point $(1.1, 0.8)$ in the middle of the rectangle. Furthermore, in both cases, we fix the surface along the two edges of the rectangle situated on the coordinate axes, and only here.

We note that the intensity remains fixed along the boundary and use the method of characteristic, as described in [3] for the solvability study, to find the numerical solutions too. In this classical method, theoretically the solution surface is obtained as the union of the characteristics, and as long as the projections of the characteristics don't cross, this gives a smooth solution. In practice, one compute a sufficiently fine grid of characteristics and interpolate in between. If some of the characteristics cross, the solution surface will get creases and self-intersections.

**Remark 1** *This is not the numerical method used in our actual implementation. It is used here since it is superior to give a good insight into the problem.*

In the first example, $c = 0.14$. Fig. 1 shows the characteristics and Fig. 2 their projection into the plane. Apart from two parts of characteristics that are appearing by re-entering the domain and thus not affecting the solution, these characteristics don't cross, and a smooth solution is obtained.

The second example is obtained by increasing $c$, from $c = 0.14$ to $c = 0.20$. The characteristics emanating from that part of the boundary of the original surface lying over the $x$-axes is shown in Fig. 3 and their projections into the plane in Fig. 4. As turns out from Fig. 4, the projections of some of the characteristics cross and Fig. 3 suggest a fold. To study its behaviour more closely, the sections of the surface with the planes $x = 2.0$ and $x = 2.2$ are computed and displayed in Fig. 5. The curve with the smaller crease corresponds to the intersection curve

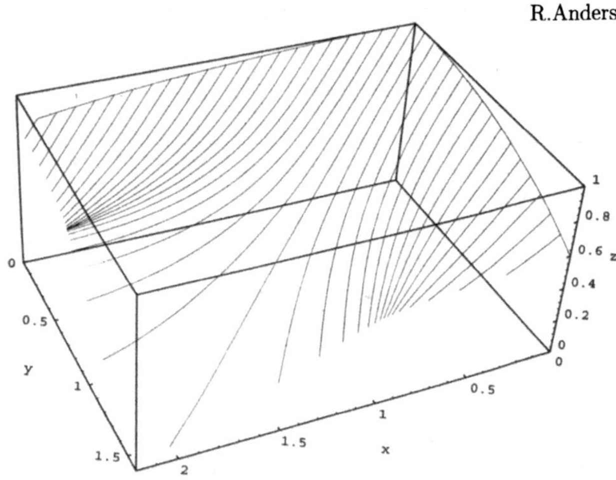


Fig. 1 Characteristics for c = 0.14.

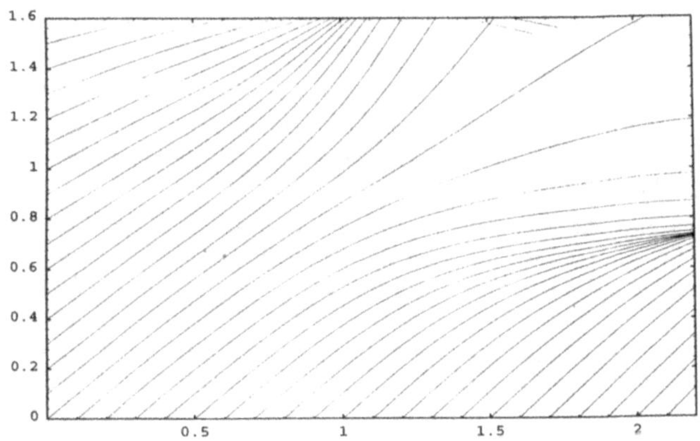


Fig. 2 Projections of the characteristics for c = 0.14.

obtained for $x = 2.0$. It reveals a self-intersecting surface with sharp creases, totally useless for any practical purpose. The dashed curves in Fig. 5 are the intersections between the two planes and the original surface and the dots mark the intersection between the characteristics and these planes.

A similar but worse result (not displayed) is obtained if $c = 0.14$ is kept, but the intensity perturbation $J$ is replaced with $J^3$, having the same maximum as $J$, but with a bigger $C^2$-norm, indicating that this norm is the proper size measure, not the maxnorm.

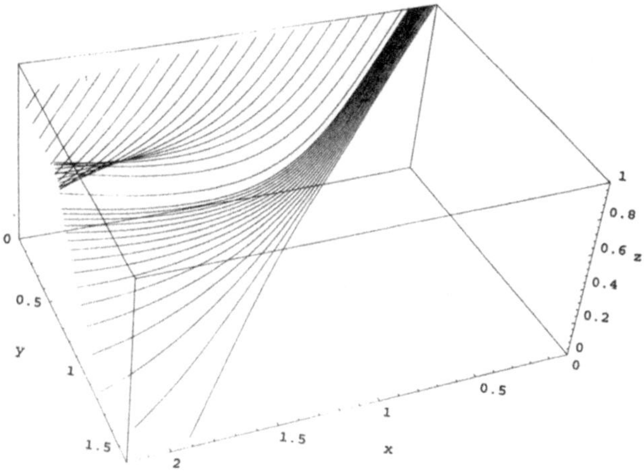


Fig. 3 Characteristics for c = 0.20.

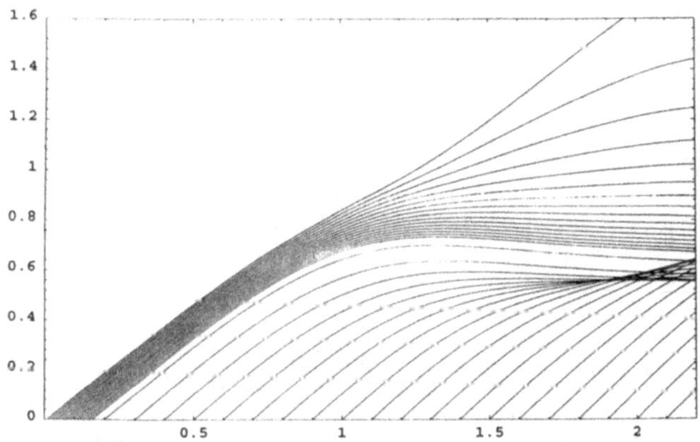


Fig. 4 Projections of the characteristics for c = 0.20.

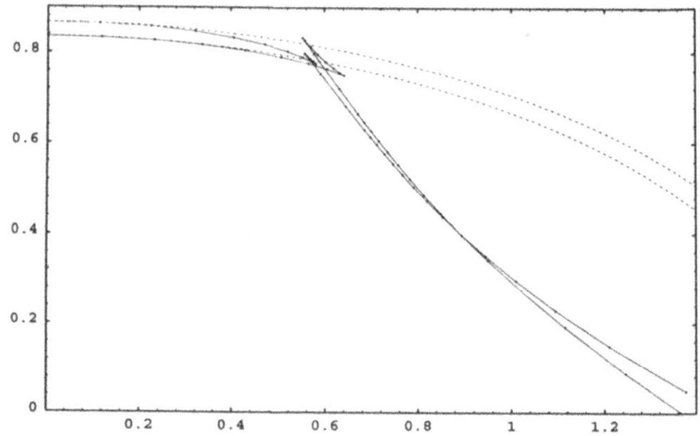


Fig. 5 Part of the surface obtained and its sections with the planes x = 2.1 and x = 2.2 for c = 0.20.

# 4 Mean and Gauss Curvature

We introduce the notation $H(u)$ and $K(u)$ for the mean- and Gauss curvature respectively, referring to [5] for their definitions and meaning. Furthermore, $\partial\Omega$ denotes the boundary of the domain and $v\mid_{\partial\Omega}= u\mid_{\partial\Omega}$ means that the functions $v$ and $u$ agree on $\partial\Omega$. The spaces $C^k(\overline{\Omega})$ were introduced in the preceding section. Here, we must consider certain subspaces instead. These subspaces are defined as follows, where $0 < \alpha \leq 1$. Let

$$H_\alpha(v) = \sup \frac{\mid v(x_1,y_1) - v(x_2,y_2)\mid}{[(x_1-x_2)^2+(y_1-y_2)^2]^{\alpha/2}},$$

where the supremum is taken over all $(x_1,y_1) \in \overline{\Omega}$, $(x_2,y_2) \in \overline{\Omega}$ with $(x_1,y_1) \neq (x_2,y_2)$. A function $v$ is said to satisfy a *Hölder condition with exponent* $\alpha$ in $\overline{\Omega}$ if $H_\alpha(v)$ is finite. For each non-negative integer $k$, $C^{k,\alpha}(\overline{\Omega})$ consists of all functions $u \in C^k(\overline{\Omega})$ for which all partial derivatives of order $k$ satisfy a Hölder condition with exponent $\alpha$ in $\overline{\Omega}$. In $C^{k,\alpha}(\overline{\Omega})$ the size of a function $u$ is measured by

$$\| u \|_{C^{k,\alpha}(\overline{\Omega})} = \| u \|_{C^k(\overline{\Omega})} + \max_{p\geq 0, q\geq 0, p+q=k} H_\alpha\left(\frac{\partial^{p+q}}{\partial x^p \partial y^q} u\right)$$

The precise statements are given by the following two theorems. Then follows a result ensuring the convergence towards the solution is quite rapid. To our experience from automotive design, one gets sufficient close to the desired curvature in at most 5 iterations and usually, only 3 iterations are enough. The results of this section was established in [1].

**Theorem 1** *Let $\Omega$ be a bounded domain in $\mathbf{R}^2$ with boundary $\partial\Omega$ in $C^{2,\alpha}$ and let $u \in C^{2,\alpha}(\overline{\Omega})$. If $f \in C^{0,\alpha}(\overline{\Omega})$ and $\|H(u) - f\|_{C^{0,\alpha}(\overline{\Omega})}$ is sufficiently small, then there is a unique $v \in C^{2,\alpha}(\overline{\Omega})$ satisfying $v\mid_{\partial\Omega}= u\mid_{\partial\Omega}$ and $H(v) = f$.*

**Theorem 2** *Let $\Omega$ be a bounded convex domain in $\mathbf{R}^2$ with boundary $\partial\Omega$ in $C^{2,\alpha}$ and let $u \in C^{2,\alpha}(\overline{\Omega})$ be strictly convex on $\overline{\Omega}$. If $f \in C^{0,\alpha}(\overline{\Omega})$ and $\|K(u) - f\|_{C^{0,\alpha}(\overline{\Omega})}$ is sufficiently small, then there is a unique $v \in C^{2,\alpha}(\overline{\Omega})$ satisfying $v\mid_{\partial\Omega} = u\mid_{\partial\Omega}$ and $K(v) = f$.*

**Theorem 3** *In each of the immediately preceding theorems, there is a sequence $(v_i)_{i=1}^{\infty}$ in $C^{2,\alpha}(\overline{\Omega})$, with each $v_i$ the solution of the linear, elliptic boundary value problem obtained through linearization of the curvature problem, converging quadratically in $C^{2,\alpha}(\overline{\Omega})$ to the solution.*

# 5 Normal Curvature and 'Sectional' Curvature

Since we will only consider graphs in this paper, we may specify the direction associated with normal curvature, see [5], simply by a vector in the plane over which the graph is defined. Normally, this direction will vary with the point in which we are considering the normal curvature. Since the length of a vector specifying a direction is irrelevant, we will assume each one to be of unit length. Thus we are led to consider unit vectors $\mathbf{w}(x,y)$ for $(x,y)$ belonging to some domain $\Omega_0$ in the plane. Often, we will describe $\mathbf{w}(x,y)$ through its components $\xi(x,y)$ and $\eta(x,y)$, i.e. $\mathbf{w}(x,y) = (\xi(x,y), \eta(x,y))$.

We will assume $\Omega_0$ is simply connected, "a domain without holes", that contains the closure $\overline{\Omega}$ of the domain over which the graph is defined and that the component functions $\xi$, $\eta$ are smooth. More precisely, we assume that $\xi, \eta \in C^k(\Omega_0)$. From a practical point of view, it is usually easy to find such an $\Omega_0$ and supply the user with convenient means to generate the required vector field with sufficient accuracy. A simple example is given in Fig. 5 where $\Omega_0$ is simply a rectangle containing $\overline{\Omega}$ and the vector field is obtained by bilinear interpolation of the four user-defined unit vectors at the vertices.

Associated with the vector field $\mathbf{w}$ are its *trajectories*, that will play a decisive role in this kind of curvature problems. They are the plane, parametric curves obtained as solutions to the initial problems

$$\begin{cases} x'(t) = \xi(x(t), y(t)), & x(0) = p \\ y'(t) = \eta(x(t), y(t)), & y(0) = q \end{cases}$$

i.e. the curves passing through a given point $(p,q)$ in $\Omega_0$, having the vector $\mathbf{w}(x(t), y(t))$ as tangent vector at the point $(x(t), y(t))$. A few of them are illustrated in Fig 5.

Being smooth plane parametric curves, these curves have a *(signed) curvature* $k$ which, since $\xi^2 + \eta^2 = 1$, is simple given by $k(t) = x'(t)y''(t) - x''(t)y'(t)$. If $k(t_0) = 0$, then the curve is said to have a point of inflection at the point corresponding the parameter value $t_0$. The quantity $k(t)$ is a measure of the deviation of the curve from the straight line through that point, see e.g. [5], p 42.

If $\mathbf{w}$ is a vector field in $\Omega_0$, a *regular quadrilateral* in $\Omega_0$ is the closed, simply connected subset of $\Omega_0$ bounded by two trajectories to $\mathbf{w}$ and two other curves intersecting the trajectories, and such that $\mathbf{w}$ is not a tangent to the latter curves in any point. An example of a regular quadrilateral is shown in Fig. 5. The thick part of the boundary in Fig. 5 is the *parabolic boundary* $\partial_P\Omega$, the only part of the boundary we are allowed to fix. The next two theorems are proved in [2], to which we also refer for the definition of the space $C^{2+\alpha,1+\alpha/2}(\overline{\Omega})$.

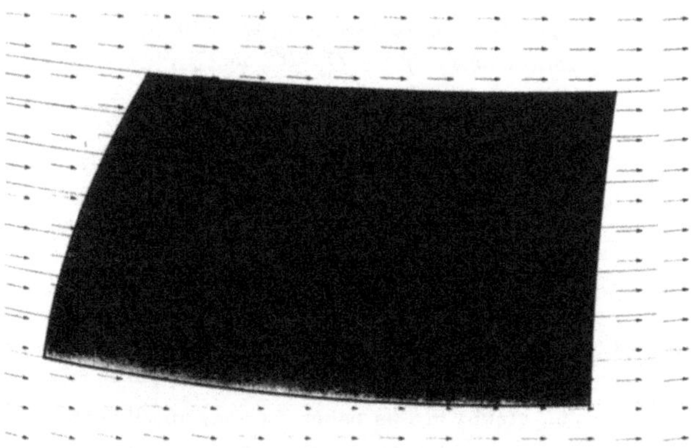

Fig. 6 Vector field, trajectories and a regular quadrilateral.

Furthermore, $s_{(\xi,\eta)}(u)$ and $k_{(\xi,\eta)}(u)$ denote the 'sectional' and normal curvature of the graph of the surface defined by $u$ respectively.

**Theorem 4** *Let* $\mathbf{w} = (\xi, \eta)$ *be a unit* $C^k$ *vector field in a simply connected domain* $\Omega_0$ *for* $k \geq 3$, *let* $\overline{\Omega}$ *be a* $C^{2,\alpha}$ *regular quadrilateral for* $\mathbf{w}$ *in* $\Omega_0$ *and let* $u \in C^{2+\alpha,1+\alpha/2}(\overline{\Omega})$ *satisfy* $s_{(\xi,\eta)}(u) > 0$. *Assume that no trajectory to* $\mathbf{w}$ *have a point of inflection within* $\overline{\Omega}$. *If* $f \in C^{\alpha,\alpha/2}(\overline{\Omega})$ *is positive and* $\| s_{(\xi,\eta)}(u) - f \|_{C^{\alpha,\alpha/2}(\overline{\Omega})}$ *is sufficiently small, then there is a unique* $v \in C^{2+\alpha,1+\alpha/2}(\overline{\Omega})$ *satisfying* $v \mid_{\partial_P\Omega} = u \mid_{\partial_P\Omega}$ *and* $s_{(\xi,\eta)}(v) = f$.

A similar result holds for the normal curvature. Here $\mathbf{w}^{\perp}$ denotes the vector field obtained from $\mathbf{w}$ by turning each vector in $\mathbf{w}$ the angle $\pi/2$ counter clockwise.

**Theorem 5** *Let* $\mathbf{w} = (\xi, \eta)$ *be a unit* $C^k$ *vector field in a simply connected domain* $\Omega_0$ *for* $k \geq 3$, *let* $\overline{\Omega}$ *be a* $C^{2,\alpha}$ *regular quadrilateral for* $\mathbf{w}$ *in* $\Omega_0$ *and let* $u \in C^{2+\alpha,1+\alpha/2}(\overline{\Omega})$ *satisfy* $k_{(\xi,\eta)}(u) > 0$. *Assume that no trajectory to* $\mathbf{w}$ *have a point of inflection within* $\overline{\Omega}$ *and that* $u$ *is monotone in the direction* $\mathbf{w}^{\perp}$. *If* $f \in C^{\alpha,\alpha/2}(\overline{\Omega})$ *is positive and* $\| k_{(\xi,\eta)}(u) - f \|_{C^{\alpha,\alpha/2}(\overline{\Omega})}$ *is sufficiently small, then there is a unique* $v \in C^{2+\alpha,1+\alpha/2}(\overline{\Omega})$ *satisfying* $v \mid_{\partial_P\Omega} = u \mid_{\partial_P\Omega}$ *and* $k_{(\xi,\eta)}(v) = f$.

**Remark 2** *In the latter theorem, the hypothesis may be somewhat weakened. If e.g. the curvatures of trajectories are identically zero somewhere, this may be compensated for by a* strictly *monotone* $u$ *cross the vector field.*

# References

[1] Andersson, R.K.E.: *Surfaces with prescribed curvature I.* Computer Aided Geometric Design **10** (1993), 431–452.

[2] Andersson, R.K.E.: *Surfaces with prescribed curvature II.* Report 1995:4, Preprint Mid Sweden University.

[3] Andersson, R.K.E.: *Smooth solvability and solutions of the image irradiance equation.* Report 1996:1, Preprint Mid Sweden University.

[4] Foley, J. D., van Dam, A., Feiner, S.K., Hughes, J.F., *Computer Graphics.* 2. Edition Addison-Wesley 1990.

[5] Hoschek, J., Lasser, D.: *Grundlagen der geometrischen Datenverarbeitung.* 2. Auflage Teubner 1992.

[6] Lions, P.L., Rouy, E. and Tourin, A.: *Shape-from-Shading, viscosity solutions and edges.* Numer. Math. **64** (1993), 323–353.

## Author

Prof. Dr. Roger Andersson
Dept. of Technique and Natural Science
Midsweden University
ÖSTERSUND
Schweden
E–mail: mg17813@gaia.swipnet.se

# The multi facets of Reverse Engineering

*Olivier Bellart*

Dassault Systems

## Introduction

The purpose of this paper is to propose an extended view on reverse engineering. For most of us, these words evokes immediately creation of digital data from physical mock-up. We suggest to generalise the concept in two directions : first to consider others media than only physical and digital mock-up; second to explore the dimension of the process flow and to study reverse engineering as a general mean for the management of concurrent processes.

We will propose some illustrations of this approach through different tools provide by CATIA current or up-coming versions. That shows, in a concrete way, that we are not discussing about dreams but on present or next future of CAD end-users.

## 1 Proposal for a extended definition

Let us assume that a DATA1 is transformed in a DATA2 through a known process. Whatever the nature of the DATA1 and the DATA2, we can say that both are representing a same object or a same part of an object. The purpose of reverse engineering is to retrieve the information DATA1 from the DATA2. DATA1 and DATA2 are an image of each other . The challenge of reverse engineering is to make all these data consistent together . In other words, if DATA1b is the result of reverse engineering on DATA2, the goal is to minimise the deviation between DATA1b and DATA 1.

## 2 PMU/DMU correlation

The best known reverse engineering process is the correlation between Physical Mock-Up and Digital Mock-Up. In this area, we can see an evolution in the processes of most of automotive and consumer goods companies. In the previous situation the PMU is the leader. All main decisions are taken on the physical model; it is the reference and the CAD data is seen as a powerful tool to create physical data thanks to digitised data processing and milling. The process have been improved and the Digital mock-up is now the reference for the product validation . Thanks to advanced tools as realistic rendering, it is possible to check on the screen the shape quality . That allows to decrease the number of physical models, avoiding some accuracy problems and minimising costs. Physical models are now simple tools for modelling. After manual modifications , digitisation is done and the cloud of points is processed and transformed into geometry in order to update the CAD model.

The data accuracy required today in industrial processes and the new digitisation technologies as laser require powerful tools for large amount of scattered points management and surfacing data creation. Dassault Systemes and Imageware , world leader in digitised data processing joined their skills and technologies to realise CATIA Cloud to Geometry product .This product offers a complete set of tools to create, manage and process large cloud of points with seamless integration with downstream applications.

*Illustration cloud*

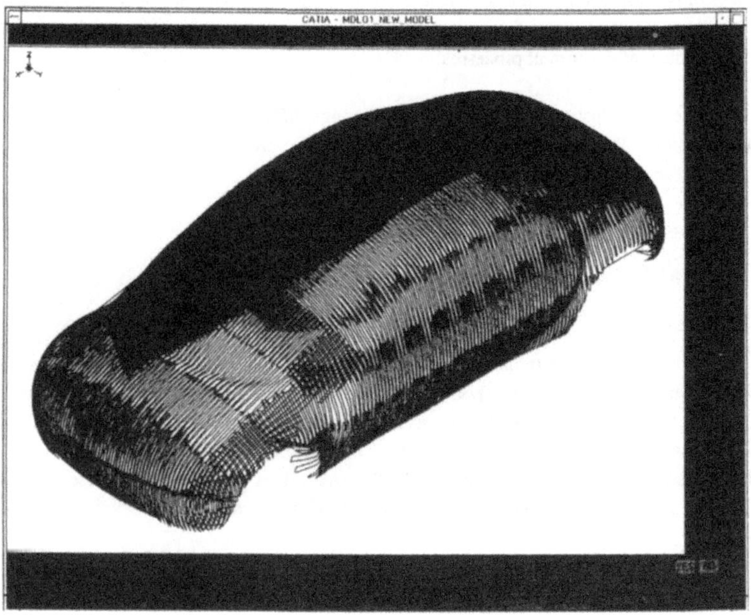


Figure 1: CATIA Cloud to Geometry

# 3 Create geometry from pixel information

Digital mock-up and Physical models are not the only media available. One common way to represent an object is Picture data. It is easy to do ( manual drawing + scan or paintbox ), it is intuitive ( no skill is required). So, picture data represent a big potential information. The usual link between digital geometrical data and picture is the computation of realistic rendered images thanks to ray tracing technologies. The result is 2D when the initial data is 3D. So there is obviously some lacks of information that prevent from a easy direct reverse engineering of the picture into 3D geometrical data.

However, with the help of additional information or assumption, extraction of geometry from picture appears as a very promising and user-friendly process.

One first significant step is to make elevation of drawings. Dassault Systemes has developed technologies to create in one operation 3D complex logos from simple picture file. This single operation includes image filtering, contour detection, elevation, draft angle creation , wrapping on free form surface and topology creation. So, user can create, in a minute, complex 3D logos, shoe or tyre prints, part identifier or others features, from a simple drawing. Options allow him to manage the multi-levels depending on the colour and to choose the draft angle and the section type: linear or curved . The created geometry can be included immediately in complex CATIA design . Through this process, modification of the geometry can be done through an simple drawing update.

*Illustration Engraving*

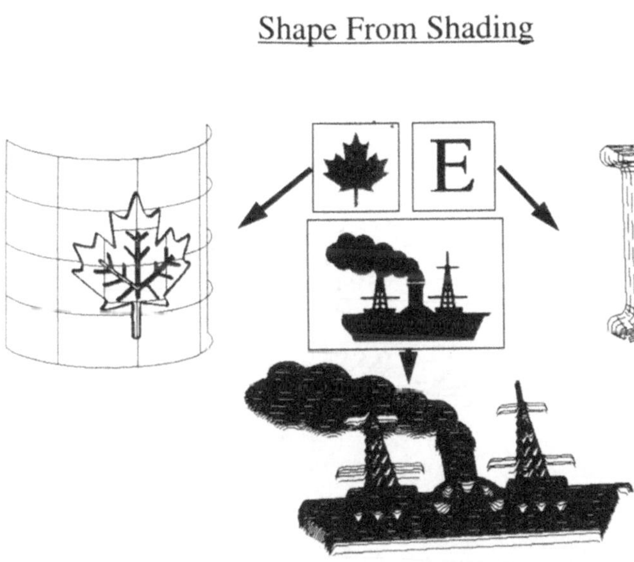


Figure 2: Examples of creation of 3D geometry from pixel information

# 4 Take into account physical constraints.

The reverse engineering concept can be extended through the complete product process as well. The data can be converted from one medium to another. it can also be transformed along the different stages . A same part can have different views depending on the step in the process. For instance, the geometry of the tools is not the same than the geometry of the final part. Physical effects as springback leads to different geometrical shapes. If we assume that for a given tool, we are able to compute the shape deformation, a reverse engineering process is needed in order to compute from the final shape the geometry of the tool that will leads to the desired final geometry. The need here is to have powerful capabilities than allows to deform globally extremely complex shapes, including outer shape, inner structures and all connecting geometry in a consistent way, keeping unchanged the character of the shape and preserving all desired continuities. One additional requirement is to avoid data size explosion.

Dassault Systemes will provided in upcoming version tools to address this reverse engineering process . User can define global space deformation from technological information and apply it to any CATIA data , simple surface or curve, as well as complex sets of surfaces or volume.

*illustration global*

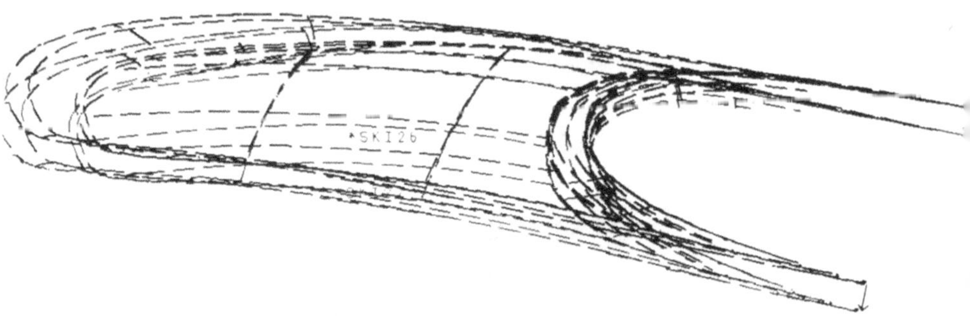

Figure 3: Example of a hood springback compensation

## 5 Digital mock-up as repository allows integration of concurrent engineering processes thanks to multi reverse engineering technics.

All processes along product definition are more and more concurrent. That makes critical the data flow between the different stages and the different views of the product. The synergy between the different processes must be maximised in order to reach the business goals. The possibility to exchange and convert data through advanced reverse engineering processes is key for general efficiency.

The integration of all the technologies involved in the product definition requires a single repository for various information as shape data, picture data, packaging data, NC data, rules, electricity data, FEM data, documentation and so on. This repository is the digital model .All these pieces of information constitute the virtual product model and the reverse engineering tools are a key component of it.

## Author

Olivier Bellart
Dassault Systems
F–92156 Suresnes CEDEX
Tel.: ++33–1 40 99 40 99 Fax.: ++33–1 42 04 45 81

# Structural Optimization and its Influence on the Design Process

*Ingo Raasch*

BMW AG München

## 1. The Changing Design Process

### 1.1. The Design Genius

The design process in the past was simple and easy to understand. The designer knew the necessary developments, the function of the design, and the necessary requirements. The design communication was solely the responsibility of the designer. This design process was never subdivided into different tasks. Its sequence was never analyzed. This process is known today as simultaneous engineering. The ingredients of a good designer were: mastering of drafting, practical experience, and use of standard references (fig 1). That meant recognizing the similarities between the new and existing designs in order to capture the important phenomena which governed the design functionality. Knowing these phenomena, simple models could be constructed, which described the essential design functionality . Reference books like Hütte, Dubbel[1], or Marks describe many

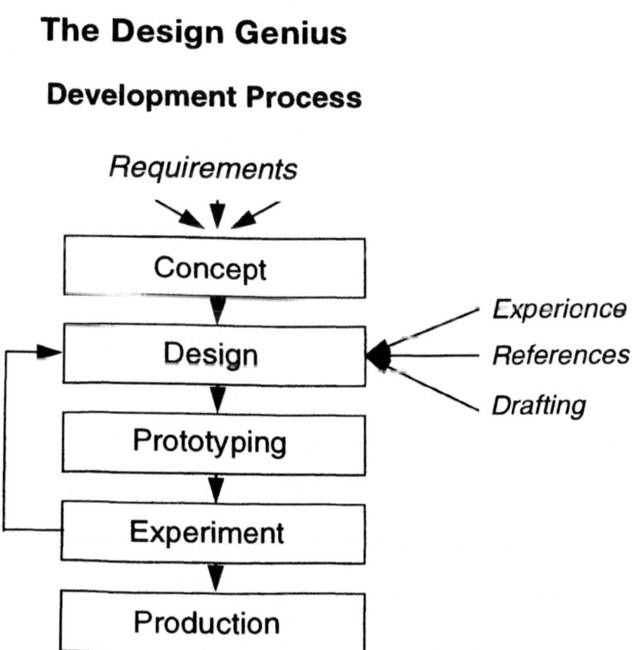


Fig. 1

phenomena, showing their influencing parameter, and how to analyze them. The understanding of the governing phenomena very often influenced the design much more than performing any model calculations. Recognizing the existence of an important phenomenon in the design distinguished a good from an average designer. Historically, the designer was responsible for the geometry and the function of the design. Prototyping and experimenting only had to prove the function of the design before the start of production.

### 1.2. The new Profession: CAE-Engineer

As the design environment became more complex, the designer realized that these simple models were no longer adequate. At the same time structural analysis techniques matured. However, these analyses were no longer under the control of the designer, but rather spawned a new profession of analysts or CAE engineers. This necessary division of labor in geometry and functionality had some major draw backs; communication and complexity are the most prominent. The vocabulary of the practical designer and the theoretical analyst had very few words in common, but even these had different meanings for each of them. The other problem was the question of interface between the two groups. What is the necessary geometric definition for the analyst and what is the best way to describe the required functionality to a designer? The current solution requires the designer to supply the complete CAD data to the analyst. The analyst reduces the complete CAD data to only such data that are relevant for the analysis. However, the question is raised: what is relevant? In order to be on the safe side, the analyst defines relevant to be all the details that can be accommodated by the available computing environment. This approach ensures accountability for all possible phenomena . Usually, the CAE software understands the CAD data to the extent that the CAE engineers understand CAD, which typically is not very well. This means that the redefinition of the original design for the analysis software is treacherous, labor-intensive, and time consuming. The definition of the design working environment has to be carefully considered. Finally, the amount of detailed results is overwhelming and the proof of functionality is not always trivial and by no means automatic. It is necessary to isolate the design governing phenomena in the final results. The generation of geometry and the proof of functionality are two steps in a design loop. These steps have to be carried out until an optimal design is reached. However, in practice these two steps are performed only once. The consequence is that the analysis must prove functionality since the CAD data are already complete and the prototyping has begun.

### 1.3. Back to Simultaneous Engineering (back to the future!)

It is obvious that this development process needs to be improved. The goal must be to reduce the disadvantages of the division of labor into two sequential steps. In order to solve this problem in the sense of simultaneous engineering, the designer and analyst must work closely and simultaneously on the same design. The designer defines only the geometry necessary for the analysis and the analyst creates the required analysis input as quickly as possible. This means that design and analysis are once again a single step, but are carried out by the team designer and analyst.

One major obstacle in this cooperation is the long time needed by the analyst to create the first analysis model. The time consuming tasks are the preparation of the CAD data such that the Finite Element Preprocessor can understand it, as well as the meshing process itself. It is astonishing that tremendous amount of man-hours are still needed to create the Finite Element mesh, since both CAD and FEM have been used for over 20 years and it was clear from the beginning that the interface between them would be very important. The creation of the first mesh is very time consuming, but variations of a given design can be analyzed very quickly and efficiently. The necessary computing power is easily provided by today's hardware. Therefore, the most significant analysis advantage can be achieved if it is used in a design loop in order to improve the design. Only proving functionality is very expensive and time consuming.

In a development environment, where the new product is a refinement of an existing product, it is advantageous to save the FEM-simplified CAD data and the meshing strategy. Subsequently, for the new product design, only the changed parts need to be created or modified from the new CAD data. Nevertheless, in the future it will still be necessary to generate the FEM-CAD data automatically.

When the designer and analyst are involved together in the design process, complexity of the design will increase after each iteration. The expectations of the analysis results will change from qualitative to quantitative. In other words,. the design process starts with a simplified analysis model, which will increase in complexity after each iteration. These facts imply that the analysis tools should be used in a development of a design concept.

Topology optimization is emerging as an additional analysis tool, which can offer suggestions for the location of. load carrying members in a given design space ( 2).

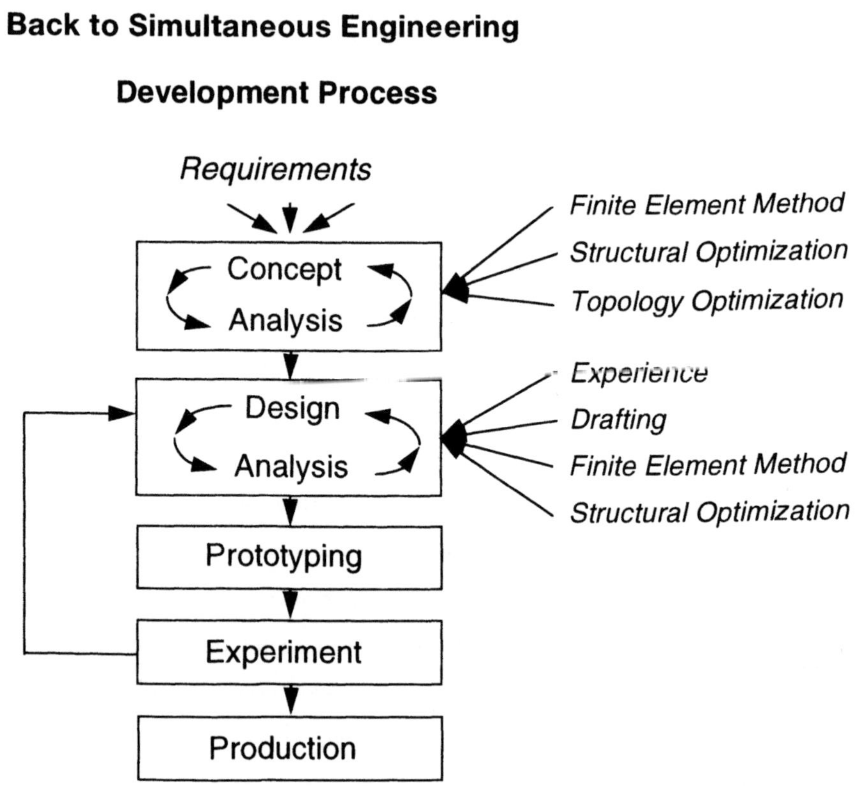


Fig. 2

## 2. Optimization

Whereas the interface problems are essentially still unresolved, the analysis tools have been improved in many ways. One very promising improvement is the emerging possibilities of structural optimization. Before discussing the possibilities of optimization further, it must be clearly stated that the usage of optimization will not reduce the development time, but rather it will improve the functionality of the product. Also, the development process must be a looping process such that the analysis has the possibility of influencing the design. The analysis requires that the operational environment be described. Optimization requires that this description is complete, since during the optimization process the relative importance of certain loading conditions can change dramatically. Furthermore, the design variables represent a new type of data and one has to prescribe their bounds and interdependencies.

Structural optimization is subdivided into three disciplines: property optimization, or sizing, shape optimization, and topology optimization.

### 2.1. Property Optimization

#### 2.1.1 Shell thickness

Property optimization requires analysis models that are reduced in one or more dimensions by using more abstract mechanical models such as shells or beams. CAD uses a similar abstraction: sheet metal structures are usually represented by one of its surfaces and the thickness is a separate item, often a simple comment in the CAD data. Because of this special role and its independence from the remainder of the CAD data, shell thicknesses are an ideal item for improving the functionality of a design using CAE. This approach avoids conflicts with designers who generally do not like outside interference with their designs. The other advantage of shell thickness optimization is that it can be done rather late in the development process and the exchange of data between CAE and CAD is minor and fairly simple.

Following this theoretical consideration, an example from the car product design will be provided [3]. In this example, the number of constraints is almost overwhelming. However, the thicknesses of the car body panels can still be changed fairly late in the development. Conversely, the design usually contains the experience of a large design team and it should be rather close to a practical optimum. Despite this fact, optimization still can achieve considerable weight savings. For the optimization analysis the finite element model represents one-half the trimmed car body with approximately 140.000 degree of freedom (DOF). There are 5 static and 2 dynamic loading conditions with 4 different boundary conditions. The static loadings define the static stiffnesses rather than true operating loads. The dynamic loads are used to define the global vibration conditions. The initial car body configuration fulfilled all the static and dynamic stiffness requirements. The objective was to lower the overall weight but keep the same stiffnesses. The body panels had practical constraints imposed such as: the thickness should not be less than 0.7 mm and not more than 2.5 mm. After approximately 10 hours of analysis on a very fast supercomputer, the result showed a theoretical improvement of 14 kg. With these recommendations the designers were able to save almost 10 kg in the actual car body.

This exercise not only reduced the car body weight, but also increased the designer's experience such that the analyst could not reach the same weight reduction with the next car. Nevertheless, panel thickness optimization is a regular step in the production design today .

### 2.1.2. Beam Properties

In the early development of structural analysis, beam structures played a dominant role. In shell structures one of their dimensions is replaced by a more abstract mechanical model. The beam is a mechanical model in which two of its dimensions are replaced by more complex properties. The geometry of the beam is represented by its neutral axis, which is not shown in standard CAD data. The advantage of the beam model is its compactness. Also, there are many engineers who have a very good feeling for beam behavior. It also has many disadvantages, especially for inexperienced engineers. Some of these disadvantages relate to the need for: recognition of a beam in the actual structure, its behavior at its ends, joints stiffness, curvatures, and so on. Therefore, in recent years beam models have been avoided in CAE. Most structures were represented either by shell or volume models, which increased the size of the finite element models tremendously. However, there are still areas where models of beams and shells are used. The most important one is in the area of concept design. In this case, the predominate constraint is the time during which a CAE evaluation has to take place.

When using the mechanical model of a shell, the geometric item such as thickness is transformed into a shell property thickness which is more or less independent of the remaining geometry. Use of beam models has a significant influence on the geometry. The cross section of a beam is transformed into properties such as area, bending moment of inertia about two axes, and the torsional moment of inertia. The true geometry is replaced by the neutral axis of the beam. The geometry used in the analysis is quite different from the usual CAD geometry and is generally created by the CAE engineer.

In concept design, a designer and a structural analyst develop a new car structure together, starting with an analysis model and not a CAD sketch. During this process many variations in the topology are tested in order to find the proper starting point for more detailed analyses. Some highlights of this process will be shown for the E1, a BMW electric powered concept car, which will fulfill the California zero emission requirements. The car body is a frame structure of aluminum profiles and is enclosed by plastic sheets . The initial analysis model is made of box beams and shells. Even if the true cross section is not a box beam, it was approximated as such. The finite element code formulates the beam only by its neutral axis and properties. The finite element preprocessor has the cross sectional geometry information and can present it to give a better impression of the true spatial design configuration. The analysis model has to be fairly small since the life span of a design idea is in the order of hours, and

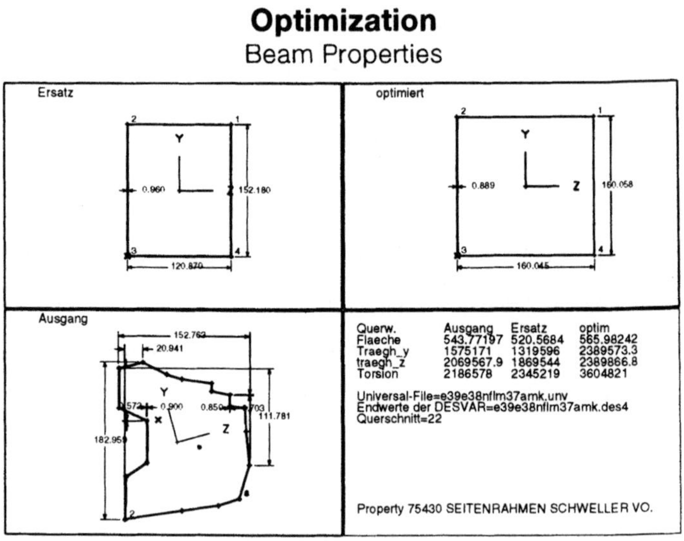


Fig. 3

the evaluation of this idea has to be realized within its life span. After a certain topology has been selected, the cross sections of the beams are adjusted by optimization. It is assumed that the cross section will be a box beam with height, width, and two different wall thicknesses, in order to give a designer a good indication of cross section size and necessary properties (fig 3). After the designer has created the cross sections, taking into account all other design constraints, the analysis model is updated with the new beam properties. A second optimization is carried out using only the wall thickness of the beam profiles as design variables. Other properties of thin walled cross sections are approximately linearly dependent on the wall thickness.

The analysis model is only one-half of the symmetric structure. The loading is static with different boundary conditions and loads superposition. Figure 4 shows a short history of the concept design with its major topology changes and a number of optimization calculations, which were carried out in each design stage. From this figure it is clear that optimization is only a tool. It is a very important tool, but it does not replace the designer's ingenuity. Since the optimizer finds local optima only, it essentially improves a good design, but does not find one.

After these initial design steps, the geometry has to be created in CAD. If the analysis model is generated directly with finite elements, then there is no geometry that can be transferred to CAD. However, in most cases some analysis geometry is generated in the preprocessor which can be transferred to CAD and used as sketches.

**Optimization in Concept Design**

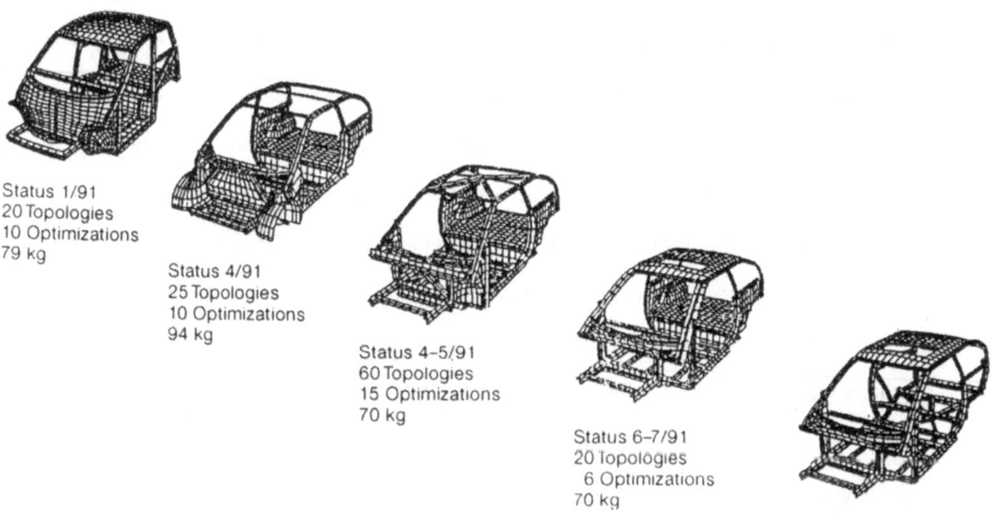


Fig.4 Example E1 - Development History

**2.2 Shape Optimization**

In property optimization some geometric items are transformed into a property, which can be changed independently from the remainder of the geometry. In beam models the shape of the cross section is optimized, which is a very special case of shape optimization. In this section that case will not be considered, but rather a more general approach. Part of a structure surface can be geometrically changed in order to improve the design. For the analysis it is not only necessary to obtain the geometric description of the structure, but also its allowable variation. The ideal situation would be a close coupling of CAD, finite element preprocessing, and analysis. Traditionally these were three independent business fields for software companies. This led to a

lack of communication between different competing companies with the effect that their interfaces were often good filters or isolators. Therefore, the first steps in shape optimization were taken by creating a limited CAD functionality in the analysis tool.
At BMW, shape optimization was started with the connecting rod [2]. This type of optimization can be very involved and complicated. The loads on the structure, various constraints, and the definition of the shape variation, which is usually a superposition of various user defined shapes, are needed. The optimization analysis will determine the optimal scaling for each defined shape. The definition of these shapes is time consuming and needs significant experience, since the shapes must be able to influence the constraints and the objective in order to obtain a meaningful result.
The example of shape optimization is a crankshaft that was designed as a forging. In order to save cost, it had to be produced as casting with the same dynamic stiffness. Figure 5 shows the original steel model. The model is simplified for the optimization and is carefully meshed, such that it can sustain large mesh distortions. The analysis model is a half segment between the main and piston bearings. The analysis involves statics and modes for two different boundary conditions. There are a total of 22 trial

**Shape Optimization**

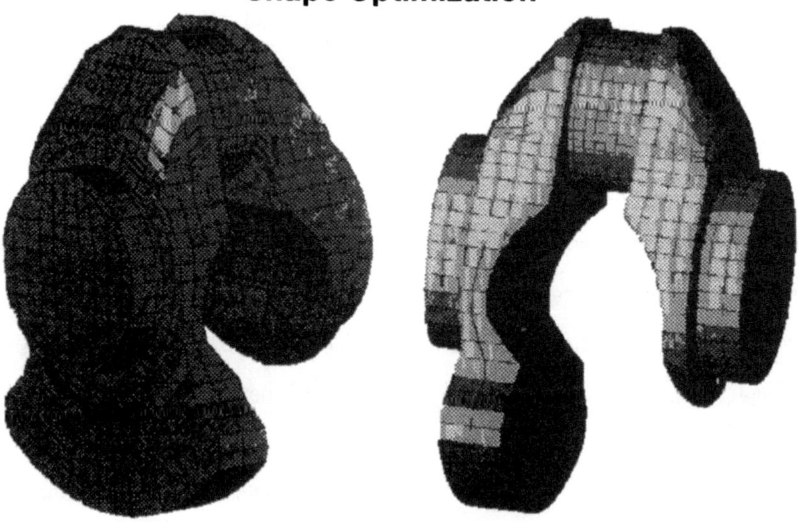

FEM-Model of Steel Crankshaft

Fig. 5

shapes on the free surface of the segment. The objective function is a combination of

**Shape Optimization**

Iteration Steps

Fig. 6

weight minimization and eigenvalue maximization. The optimization was started in the infeasable region. Figure 6 shows some intermediate steps in the iteration history. After 15 iterations all constraints were satisfied. The new design reduced the production cost by approximately 40%.

Topology Optimization

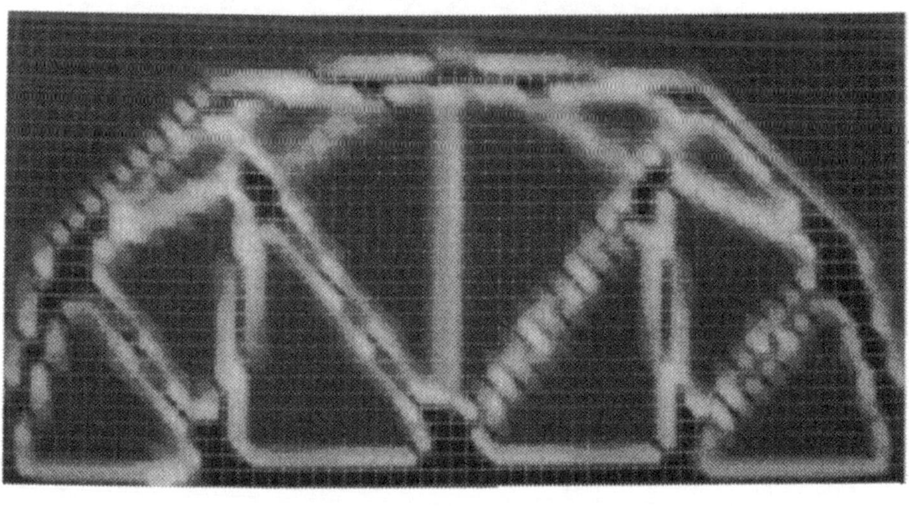

Solution of a „Bridge Type Loading"

Fig.7

With the inevitable mergers in the CAD/CAE software business, this procedure can be done differently today. In the CAD system a parametric model of the crankshaft can be developed. The internal analysis and optimization code can take the CAD parameters as design or shape variables and derive an optimal CAD raw design. The raw design can then be refined for the actual production. The difficulty for the user is to define a proper parametric model that will allow a great variety of shapes.

### 2.3. Topology Optimization

In concept design the necessary topology optimization is done by trial and error. However, there are some recent developments which allow topology optimization [4]. It will not generate a true design, but will derive suggestions where it might be advisable

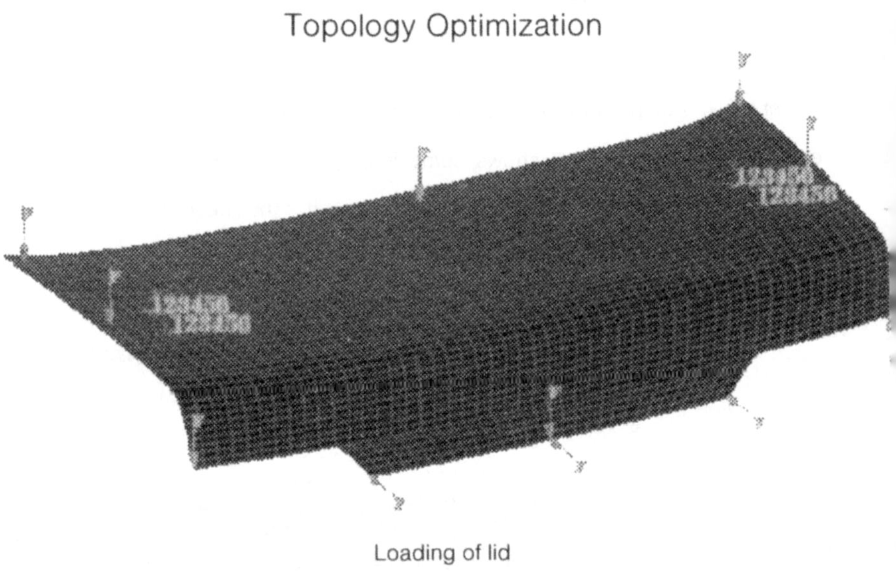


Fig. 8

to concentrate or take away material. The showcase example is a framework bridge (fig 7). There are two supports, left and right, and three loads in-between that must be carried. The design space is a rectangle above the supports. After some analysis iterations the procedure recommends concentrating the material in the highlighted areas. There are several academic codes that can handle examples such as this.

Topology Optimization

Optistruct Recommendation

Fig. 9

However, there are a few commercial codes which already have demonstrated some practical applications;e.g., Altair's car lid reinforcement. The skin of the lid is given and the problem is to find a proper understructure for ten operational loads (fig 8). OptiStruct proposed the design suggestion shown in fig 9. This was then transferred to a design. In a further step, this design can be optimized by a traditional shape optimization cycle.

The new geometric item in topology optimization is the design space, the space that may be available for the design. This space will not be completely filled with material. Usually 20 to 30% is used.

## 3. Optimization and its Influence on Design

For analysis and optimization in production design, the CAD data come first and need not be complete (fig 10). When generating the analysis model, the CAD data have to be filtered, and, unfortunately, often "cleaned." The result from the optimization for shell- like structures are the shell thicknesses. Today's processes must have a better interface, but no redesign of the development process is necessary.

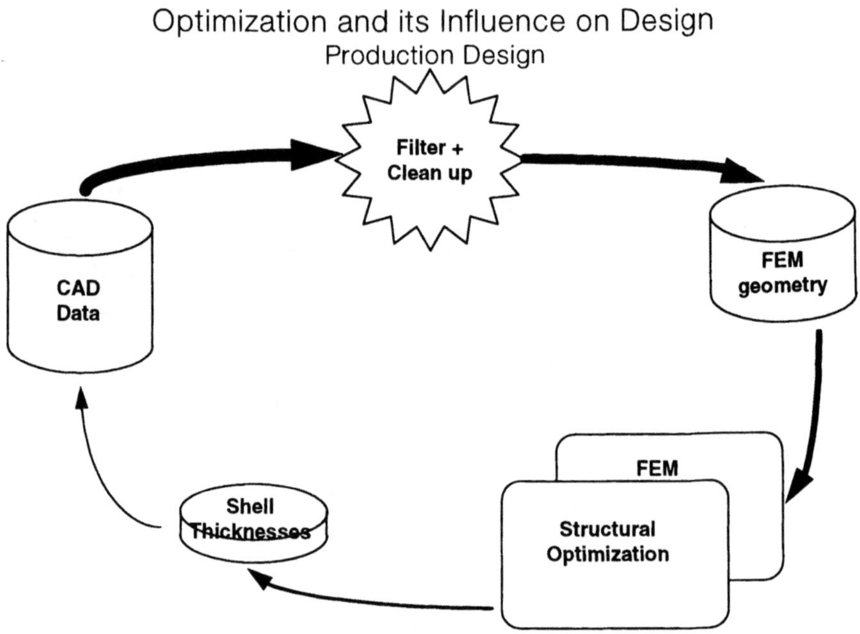
Optimization and its Influence on Design
Production Design
Filter +
Clean up
CAD
Data
FEM
geometry
FEM
Structural
Optimization
Shell
Thicknesses

Fig. 10

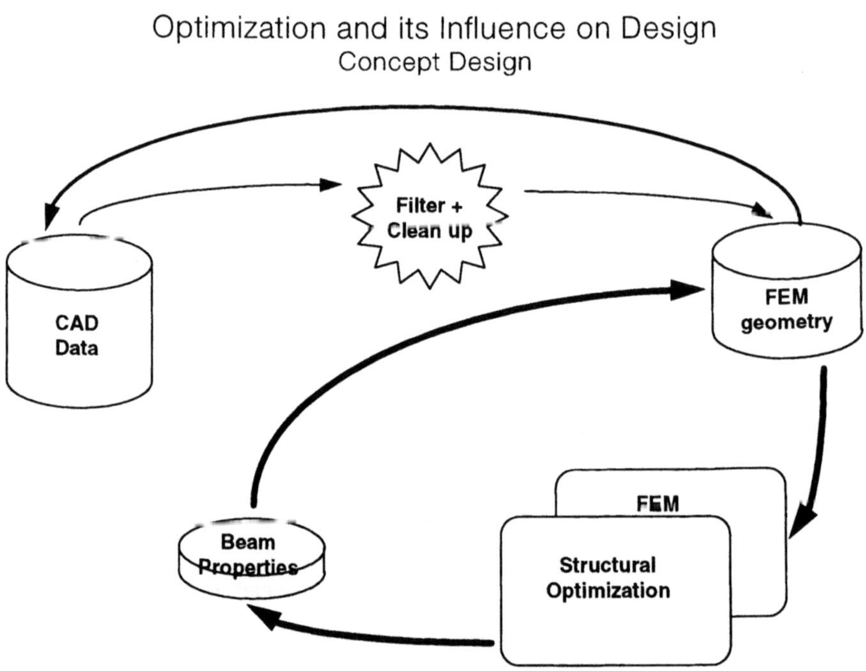
Optimization and its Influence on Design
Concept Design
Filter +
Clean up
CAD
Data
FEM
geometry
FEM
Structural
Optimization
Beam
Properties

Fig. 11

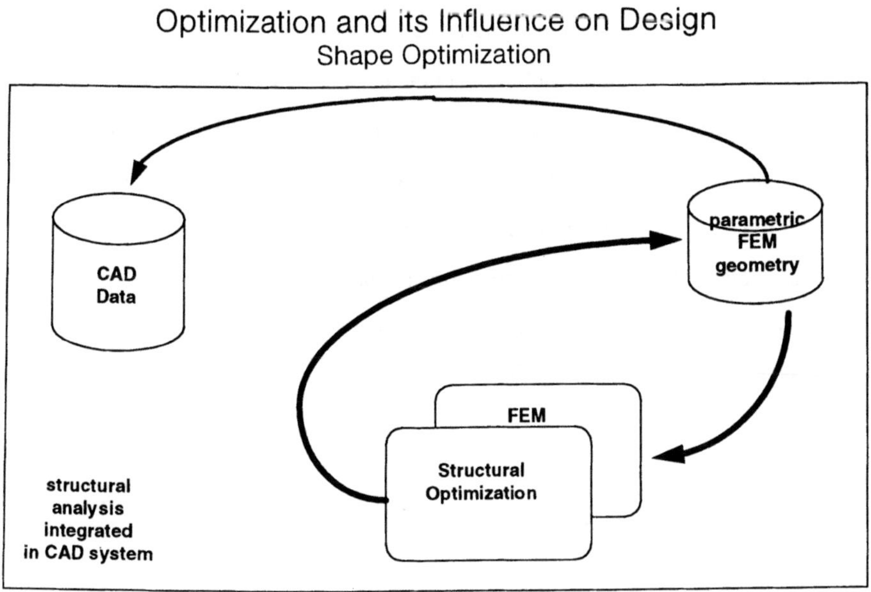


Fig. 12

In concept design, the CAD geometry and analysis geometry differ quite dramatically. Only minor parts can be transferred from one to the other(fig 11). The geometry of the load- carrying structure is first defined by the analysis and optimization. Subsequently, design geometry can be derived from the results.

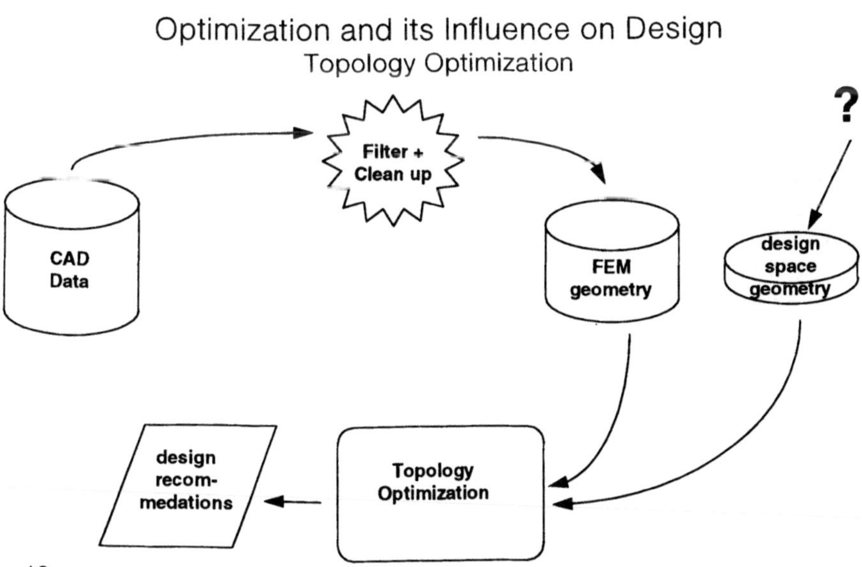


Fig. 13

In shape optimization a parametric model has to be generated, which will be used for optimization(fig 12). It is important that reparametrization can be done easily in order to give the optimization process sufficient freedom to find an optimal configuration. The design parameters will be determined by analysis. The initial geometry model is analysis- oriented, but nevertheless, is a good starting point for the designer.

Topology optimization needs analysis geometry for the boundary prescribed by system constraints, as well as the definition of the design space(fig 13). The results of topology optimization are currently not transferable to CAD.

References

[1] Dubbel - Taschenbuch für den Maschinenbau, 17. Auflage, Berlin, Springer 1990

[2] Raasch, I; Irrgang A; Shape Optimization with MSC/NASTRAN; MSC/NASTRAN User's Conference, Rome, 1988

[3] Raasch, I.; Structural Optimization with Solution 2001 in the Design Process, MSC/NASTRAN World User's Conference, Orlando, Fla., 1994

[4] --; Software produces best structure from scratch, Machine Design, p. 28, Sept. 12 1994

## Author

Ingo Raasch
BMW AG
EG-K-3
D-80788 München
Tel.: ++49(089) 382-41627 Fax.: ++49(089) 382-44745

# 3D-Shaping in CAD Systems by the Steps Creation - Inspection - Optimization

*Sabine Roth-Koch*

Fraunhofer-Institute for Manufacturing Engineering and Automation (FhG-IPA)

**Abstract:** Using fast tactile or optical sensors is a new way to manage the process of geometry definition in the CAD surface model, the 3D-shaping, without the time consuming CAD-construction of the geometry. The digitising of the shape of physical models results in 3D measuring 'point clouds'. Working on these 'point clouds' a new CAD philosophy and new CAD methods are necessary to create CAD surface models. Subject of the paper are CAD methods for feature-based definition of free formed surfaces in CAD models. These methods are used for mapping geometrical and topological information which is extracted from the measured points.

## 1 Something to think about

A product is characterised by a vast quantity of data. This is created and processed in very different ways throughout the product's development process.

Initially, information carriers are usually graphical or shape models, in other words, technical drawings or physical models which contain particular product properties. The product's desired ideal form is defined in the structure and in properties of these design or prototype models. This ideal form is then further adapted and optimised during the development stages which follow the design, until the form which suits the intended application is found. Through the use of new generative production processes, this sequence is currently being increasingly supported and altered by prototypes. This has given a new relevance to the use of physical models or prototypes in product development.

Apart from graphical or shape models, design and manufacture require a formal reflection of the constructional reality present in the respective development stages - the generation of computerised, so-called CAD models. In other words, the geometric information contained in the physical models must be formalised, in order to be able to generate product and manufacturing data on the basis of computerised models. This is usually achieved with the aid of CAD systems.

In CAD systems, computerised models are still produced even today using the same design aids and applying the same design principles used to produce technical drawings: from point coordinates and dimensions using 'ruler and compasses', which are programmed as geometric methods. These aids are not adequate for a large number of technical products, such as components from the aviation and space industry, the motor car industry, shipbuilding and also parts from the consumer goods industry, e.g. household appliance housings. This is because these products have a highly complex structural form for functional, ergonomic or stylistic reasons. What is meant by complex is that the surface is made up of both planar areas developed according to design principles, e.g. outlet radii, bores, level profiles, and also of freeform, empirically designed areas, e.g. spline curve areas on a car body. In traditional design, such freeforms were produced on the basis of discrete points using curve templates or spline battens. In geometric interpretation these points are piecewise approximated or interpolated. This is also achieved algorithmically in CAD systems. CAD systems therefore provide functions which enable

- *standard geometrical surface forms*, which permit simple analytical description in the mathematical sense, and
- *freeform surfaces*, which do not permit simple analytical description, but must be approximated as images of unknown functions, to be modelled by the computer.

A new way towards computerised modelling is being opened up by the increasingly wide use of efficient measuring systems. These mean that physical models can be digitised very rapidly and with a high degree of precision. There is a new start available to CAD modelling based on digitised information. The small amount of time required to digitise even highly complex surfaces means that the surface can be recorded as a densely described point cloud. Although the point cloud produces an extremely good approximation of the actual surface, the large volume of data generated cannot be processed by the CAD system within a realistic time. Furthermore, information on the surface is only available at discrete measuring points. Topological relationships and surface characteristics, such as natural edges or visible edges, are not taken into account. The point cloud cannot therefore replace the formal surface model, but must be regarded as the basis for its generation. There is an evident deficiency here. It is no longer a purely design process involving the discrete points and dimensions, curves and finally surfaces which is required. Instead, the important thing is to identify new methods of data preparation, in order to facilitate a 'redesign' (another frequently used term is 'surface restoration') of the physical models or prototypes.

## 2 Motivation

The aim, therefore, is to convert the point cloud into standard geometrical and freely formed CAD surfaces under design conditions.

Standard geometries are comparatively easy to produce in this case, since analytical ideal geometries exist for them, such as the cylinder for shafts or bores. This involves the use of fitting methods which produce acceptable results, if they are combined with optimisation strategies. The CAD designer simply has to specify the type of geometry and starting values for the automatic calculation within the point cloud. By contrast, a fully automatic procedure requires a geometry recognition by the system. The best possible fitting geometrical elements, e.g. cylinders or planes, are therefore fitted into the point cloud. However, this procedure only produces a satisfactory result where there are simple, continuous surface structures. A good compromise may be the combination of both processes, in other words, for the system initially to propose a geometry within a defined area which the designer could then edit and complete.

The definition of development surfaces, such as translation surfaces or rotation surfaces, and of connection surfaces, the so-called blending surfaces, is rather more on the design side and can therefore only be achieved sufficiently accurately with difficulty, if at all, by means of fitting geometries. It is advisable to continue to produce these on the basis of selected measuring points using the functions offered by the CAD systems.

This sort of 'clear-cut' procedure cannot be defined for freeform geometries. The first problem is that freeform geometries cannot be naturally or universally and particularly not geometrically classified, but, strictly speaking, can only be defined within the environment of a curve or surface point. Consequently, a series of corresponding attempts have been made to produce an approximate classification, which are based either on the application or the computerised descriptional methods. For technical applications these include, for example,

- *Aesthetic surfaces*
In this case it is chiefly the qualitative aspect in the image, the so-called fair geometry, which is decisive; frequent changes have to be made, the use of prototypes is common.
- *Generalised cylindrical surfaces*
What are important here are geometrical boundary conditions, they produce 'smooth' blending surfaces between the edges of adjacent areas with the actual course of the surfaces being unimportant.
- *Blending surfaces and fillets*
These are blending surfaces between adjacent areas whose inner course must have a particular form, the surfaces are not generally modelled, but constructed.
- *Functional or adapted surfaces (fitting)*
There is a large number of geometrical conditions, fitting into the existing point cloud takes place under approximation or interpolation conditions.

It is important to remember with this surface grouping that it cannot be clearly divided and overlaps are even possible between the classification groups. It is clear, however, that a distinction is indirectly drawn here too based on the computerised descriptional method. There is a good reason for this - because the analytical functions of freeforms are unknown, they are captured (more or less effectively) by approximation. The quality of the descriptional methods also determines the quality of the surfaces.

Which *descriptional methods* are available?

Freeform geometries are defined by adapting suitable n-degree rational or non-rational functions of the lowest possible degree within surface areas with a generally quadrilateral topology. The descriptional methods encompass approximation and interpolation algorithms. Depending on the information input, e.g. measuring points and boundary conditions, as well as the modelling aim, e.g. the desired smoothing of the measuring points, two methods are used to represent surfaces in practice:

- *interpolation / approximation* using suitable blending functions creating Tensor product surfaces, e.g. Bézier method
and
- *transfinite methods*, i.e. the joining of surface curves using suitable blending functions, e.g. COONS method.

The use of different methods implies the recognition that there is no optimum approximation function based on a suitable blending function. On the other hand, it is characterised by the wish to adapt the necessary interactions to the requirements of the design work, so that, for instance, a local or global change in the described surface can be achieved by accentuating or moving input points. The interactions required during this must be easy to execute and the results must be available in real time and be easily comprehensible. In view of the complexity of the mathematical processes, this can often only be satisfied using auxiliary solutions, e.g. with intersections.

What do the available *measuring points* look like?

In general, the following input point sets are available:

- points distributed *regularly* over the surface,

these are, for instance, points arranged in ascending order on non-intersecting polygons, such as scanlines or points on special surface curves, such as profile curves,

- points distributed *irregularly* over the surface,

these may be points which are distributed in any way or in any way within edge curves defined by points.

This information must be prepared accordingly for the description. Consequently, the description of Tensor product surfaces requires a regular control point net which must be made up of raw data with a completely random distribution. However, CAD systems do not offer the functions required to support such a procedure, e.g. the limiting, thinning out and ordering of measuring point quantities. Instead, a suitable control point net is assumed.

# 3 Prerequisites

As has already been mentioned, today the measuring techniques are well equipped for recording measuring points. Various processes which are both tactile and non-contact are used and provide sufficiently accurate and complete measuring point clouds. In order for the measuring points to be further processed into CAD models, a distinction must first be made between surface-measuring and volume-oriented digitising processes, which require different procedures. Surface-measuring processes are laser scanning or the Moiré process. Computer tomography, on the other hand, is volume-based and has hitherto been used primarily in medicine.

Direct CAD volume modelling, which does not yet have sufficient algorithmic support for freeforms, can be disregarded. Here, a so called Voxel model will be built to begin with, from which the surface model will be derived.

For direct CAD surface modelling, the following methods are available and the manipulation of these is characterised in a few words:

*Bézier method*

- is an approximation of a control point network which creates a convex hull of the surface,
- the points on the edges are interpolated,
- blending functions are the Bernstein polynomials,
- the number of control points determines the order (degree+1) of the Bernstein polynomials,
- the surface can be influenced by increasing the degree,
- only pseudo-local changes are possible in the surface.

*B-Spline method*

- is an interpolation or approximation of a control point net which forms a convex hull of the surface,

- blending functions are the B-Splines,

- the number of control points is independent of the order (degree+1) of the B-Splines,

- the surface can be influenced by increasing the degree,

- the surface can be changed locally via moved control points,

- a B-Spline curve as a piecewise defined polynomial of degree $K+1$ is generally $C^{k+2}$-continue at the transitions.

*NURBS method*

- rational Bézier and B-Spline surfaces are defined as projections of a four-dimensional Tensor product surfaces, but are not themselves Tensor product surfaces, the rational surfaces must be oriented in order to be able to use Tensor product methods,
- additional features are the weights to be specified in the control points,
- a change of weights alters the surface,
- if all weights are equal 1, non-rational surface definitions result.

In CAD systems, these methods are implemented in a variety of ways. Recently, the NURBS method has been given preference, although it is the most difficult one to survey in relation to user management and its reactions to modelling. In most cases, therefore, not all the possible parameters can be freely altered in the CAD system.

For the use of the descriptional methods listed, the following can be indicated as prerequisites for freeform modelling based on measuring points:

- the freeform surfaces are described by curves,
- they generally have quadrilateral parameter sectors,
- the randomly distributed measuring points must be arranged on control point nets and parameterised.

If the aim, therefore, is to model complex surface forms, i.e. standard forms and freeforms, interaction possibilities must be retained between the CAD surface model and the measuring data. What is necessary and advisable in this context is an adapted data model outside the CAD data structure. For the data exchange between the measuring and CAD system, the measuring data must furthermore be reduced and classified in relation to its 'significance' to the surfaces to be modelled. In this case, the use of characteristic features for freeform geometries must assume the function of the ideal geometry of standard forms. Features can be derived from quantitative and qualitative data, such as control point coordinates and smoothness, as well as from the descriptional methods, e.g. for tangential junctions. The features then contribute to the measuring point selection and to a procedure supported by algorithms and must therefore be regarded as another important prerequisite.

Inspection of the form within the constructed CAD model is divided into the actual measuring process and the subsequent evaluation of the measured values. The evaluation involves a numerical appraisal of the measured data, which is combined to produce the formal actual model. The formal models for the nominal/actual comparison can be described in various ways, for instance, using discrete points or curve or surface models. Various measuring instruments and methods are used to capture the actual data; however, all the measuring systems used supply only discrete Cartesian coordinates of surface points in a spatial coordinate system.

Another important step, the improvement of the CAD model following its inspection using a physical nominal form, is not supported by today's CAD systems. The processing of the information obtained on the actual status ends, however, with the nominal/actual comparison in the CAD system; a 'restoration' of the measured actual values in the CAD model to improve it cannot be executed. The problems associated with a 'restoration' of measured data in the CAD model is of great importance, however, and always arises when physical prototypes are produced, as prescribed by the CAD model, which are used and altered for technical inspection or optimisation.

These changes must be consistently adopted, in other words, 'retrieved', if the CAD model is to assume the role of a reference model for design cycles and, moreover, even that of a universal product data model. However, in this case the universality of the procedure is only guaranteed if the fitting of the changed surface areas is not limited to the area limitations existing in the CAD model, but can also overlap between areas.

The existing CAD system functions are therefore lacking support in four process steps:

- in the construction of initial CAD surface models on the basis of measuring point clouds,
- in the feature-based generation of nominal measuring points in the CAD model,
- in the restoration and preparation of the actual data measured on the physical model and
- in the optimisation of surface areas in the CAD model, taking account of connection conditions and continuity.

What is also important in this heterogeneous system environment is the actual use of interfaces with regard to the data transfer format produced by these interfaces. Of interest here in all applications for form definition based on measuring points is the transfer of the elements point, point sequence, point/vector sequence, polyline and the coefficients and other surface descriptional parameters.

## 3 Methods and results

The aim, this being the incorporation of new methods for the formal modelling of complex geometries on the basis of measuring point clouds, can be achieved by expanding the traditional CAD system. For freeforms, the requirement extends to their feature-based definition. In line with the peculiar nature of freeform surfaces, it is advisable to provide a sufficiently broad range of functions and to make the selection of these functions clear at a glance and comprehensible. A data preparation interface is installed in conjunction with the measuring system.

The generation of functions for the CAD system and data preparation interface must be preceded by a strict definition of the modelling aims, in other words, the question of the concrete application. In this case, it is important to be clear about whether what is needed is simply a straight forward reconstruction of the surface for the purposes of representation or filing, for a tool path or STL path calculation, for simple modifications or animation or for preparing topological information. In the other case, for construction, production, calculation or optimisation, parametrically 'true' CAD surfaces are necessary.

In the aforementioned case, that of form definition in the CAD model, the methods which form the basis of the expanded functions focus on

- the rapid *capture of the physical model surfaces* for feature-based generation of an initial CAD surface model,
- the *feature-based inspection* of this model, the comparison with the ideal form defined in the physical model and
- the *restoration of manual changes or optimisations* involving the physical model in the initial surface model for the nominal/actual comparison and its update or optimisation.

The procedure can be implemented for practical application on the basis of measuring systems currently available and using 3D CAD systems with open architectures in the data base and functions. In this case, the main part of the practical implementation is achieved through the functional expansion of the 3D CAD system and the integration of a data preparation interface with the measuring system. The main feature of the procedure achieved in this way is the creation of standard basic information on CAD surface modelling of standard and freeforms, whereby the information is reproducible with the implementation of different solution variants.

The creation of a prototype solution at the Fraunhofer Institute for Manufacturing Engineering and Automation (IPA) in Stuttgart comprises:

- a laser triangulation sensor (laser digitiser),
- a workstation for the CAD environment (CATIA) and for the data preparation interface with the measuring system (GUGS).

The main feature of the laser digitiser used is the 'synchronised scanning' technique. Here, the image size and accuracy are independent of one another, which means that a larger image field can be achieved with the same accuracy. The triangulation base is short in length, which leads to a reduction in shading effects, as well as low sensitivity to environmental light. The laser digitiser was fitted to a coordinate measuring unit. The local coordinates of the laser digitiser are transmitted via an electronic interface to the incremental path measuring system of the coordinate measuring unit. The laser digitiser can be run automatically on the three axes of the machine coordinate system. The mounting device also has four docking points, so that the position of the digitiser in the machine coordinate system can be changed to any angle position. An object can therefore be digitised from all sides.

The GUGS software which was developed at the IPA undertakes the preliminary preparation of the recorded measuring point sets in the data preparation interface. Simple surface descriptions (facetted surfaces) and path generation (section curves) can also be produced automatically here. The processed point data is delivered to the CAD system. Here, the CAD surface modelling is performed via the steps: Creation - Inspection - Optimisation using the additional functions developed at the IPA.

Compared with traditional CAD technology, a crucial improvement can be seen here in relation to the following points:

- use of *geometry primitives as meta information* between the measuring technology and CAD geometry elements (reduction in data volume),
- *feature-based test methods* and nominal/actual comparison in the CAD model,
- possibilities for the *feature-based modification* of defined surface areas (*adjustment or correction surfaces*),
- creation of a *data structure* which is *CAD-independent* and adapted to the task.

# 4 Conclusion

The work on feature-based form definition is being carried out as part of the 'Rapid Prototyping' network project started by the Fraunhofer-Gesellschaft, within the project phase dealing with the question of product modelling and reverse engineering. The preliminary results will be demonstrated in the presentation.

# 5 References

Aumann, G., Spitzmüller, K.: Computerorientierte Geometrie. Mannheim, u.a.: BI Wissenschaftsverlag, 1993, Reihe Informatik: Bd. 89

Farin, G.: Kurven und Flächen im Computer Aided Geometric Design. Braunschweig; Wiesbaden: Vieweg, 1994

Roth-Koch, S.: Merkmalsbasierte Definition von Freiformgeometrien auf der Basis räumlicher Punktwolken. Berlin, u.a.: Springer, 1996 (IPA-IAO Forschung und Praxis: Bd. 229)

# Author

Dr.-Ing. Dipl.-Math. Sabine Roth-Koch
Frauenhofer Institut
Produktionstechnik und Automatisierung
Nobelstr. 12
D-70569 Stuttgart

# Surface Reconstruction from Dense Point–Sets Using Image–Processing Techniques

*Nickolas Sapidis, Paul Besl*

General Motors Research & Development Center

**Abstract:** Current methods for extracting surface-models from "clouds of points" use standard CAGD techniques and guarantee neither *fairness* nor *closeness of fit* — important requirements for applications like Aesthetic Surface Design (Styling) and Reverse Enginnering. This paper elaborates on a new approach, combining CAGD with Image Processing, to derive surface-reconstruction algorithms that can handle dense range-images and meet strict industrial specifications.

## 1 Introduction

Approximating a dense range-image with a CAD-quality surface-model involves many problems [1] [5] [8] [11] [12] with most important

**(a)** the segmentation of the image into regions, and

**(b)** the specification of the required degrees-of-freedom in the surfaces approximating these regions.

Current techniques use primarily CAGD algorithms to reconstruct surfaces and offer only partial solutions to the above problems: Regarding **(a)**, all published methods rely on the user's expertise to segment the image, with only one exception, the work [13], where an algorithmic solution is proposed using *edge-detection.* This approach has many disadvantages as (i) it attempts to derive "surface information" from rapid changes in measurements, (ii) it assumes that any part of an image surrounded by an edge can be correctly approximated by a single surface, and (iii) it fails to identify any regions when the transition areas, between different parts of an image, have been "rounded". Regarding problem **(b)**, current techniques require the user to specify, e.g., the number of polynomial segments in B-spline curves and surfaces (to be fitted to image-data), and suggest use of postprocessors like *parameter correction* and *curve/surface fairing* to remove infidelities from the surfaces.

Image Processing has long been dealing with problems closely related to surface-reconstruction as its primary objective is the automatic extraction of shape-information from physically sensed signals (see, e.g., [1]–[4] and references therein). This field offers a wide range of techniques for dealing with issues related to transforming a "discrete (digital) surface" to a "continuous (analog) surface" and in particular it offers various solutions to both problems **(a)** and **(b)** discussed above. Of all these, we have identified *"region growing (RG)"* and *"variable-order fitting (VOF)"* as the most appropriate for producing

CAD-quality surfaces from range-images. The remainder of this paper reviews a surface-reconstruction technique, currently in use at GM, employing certain variants of RG and VOF, as well as many other tools from Image Processing. Although this procedure can handle any kind of data, its development has been targeting aesthetic-surface reconstruction from highly accurate (e.g., error $\approx 0.05 - 0.1$ mm) and very dense point-sets (typical size: $10^5 - 10^6$ points).

## 2 Dense Range Images and Surface Coherence

A variety of *sensors* is available today producing distance measurements (range images) at rates up to several millions of measurements per minute [1] [2] [3]. Most applications use a *gridded Dense Range Image (DRI)* $I = \{\mathbf{p}_{ij} \equiv (x_i,\ y_j,\ z_{ij})\}$ (see Fig. 1), where

$$x_i \;=\; a_x + s_x * i \quad \text{and} \quad y_j \;=\; a_y + s_y * j. \tag{1}$$

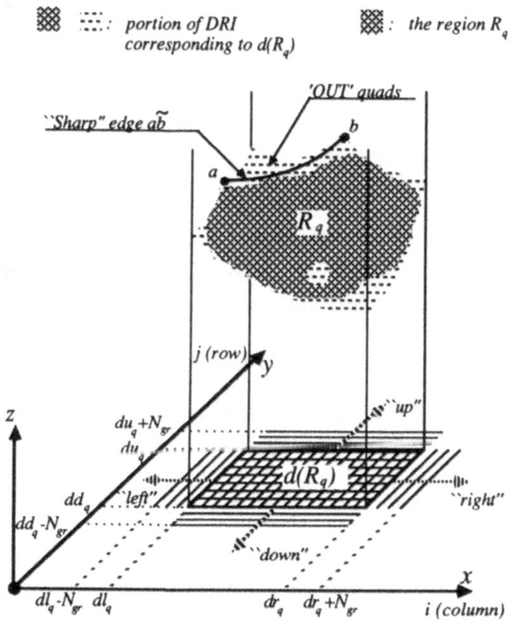


Fig. 1 A Region $R_q$, its domain $d(R_q)$, and the areas where `GrowRegionOUT` attempts to grow $R_q$.

Here, the *column number* $i$ takes values $0, 1, ...., K$, the *row number* $j$ takes values $0, 1, ...., L$, $\mathbf{a} = (a_x, a_y)$ is an offset vector, and $\mathbf{s} = (s_x, s_y)$ the increment vector. Any image location $(i, j)$ corresponding to a valid measurement $z_{ij} > 0$ produces an *object point* $\mathbf{p}_{ij}$, as opposed to a *background point* $\mathbf{p}_{kl}$ with $z_{kl} = 0$ or $z_{kl} < 0$ .

Numerous tests have confirmed that current image sensors produce highly accurate point-sets for which one can assume validity of the *surface coherence hypothesis (SCH):*

> ***the image data may be interpreted as random noisy-samples of a free-form, piecewise-smooth "physical" surface*** $z_f(x,y)$.

This hypothesis establishes the correctness of many image-processing operations [1] [2], some of which are also related to this work. These are [1] [4]:

1. *Thinning Operations:* Aim at reducing the size of a range-image by removing points encompassing the least amount of "surface information".
2. *Smoothing Operations:* Aim at removing *random noise* from an image.
3. *Normal-Vector Estimators:* These are convolution operators that produce, for each point $\mathbf{p}_{ij} \in I$, an estimation $\mathbf{n}_{ij}$ of the unit normal-vector of the "physical" surface underlying $I$.
4. *Threshold Estimators:* Produce, for the neighborhood of each point $\mathbf{p}_{ij} \in I$, an estimate of the optimal "point-to-surface error" for surface-fitting.
5. *Regions-Test in Surface Fitting:* In order for the surface $z = \zeta(x,y)$ to be an ideal fit to the points $\{(x_i, y_j, z_{ij})\}$, not only the "point-to-surface" error must be small but also $\zeta(x,y)$ must correspond to a continuously alternating *sign-of-residual*; the residual for the point $\mathbf{p}_{ij} = (x_i, y_j, z_{ij})$ is: $\zeta(x_i, y_i) - z_{ij}$.

Regarding *thinning,* it must be emphasized that, in a general purpose "point-cloud" modeling system, this should be avoided as many modeling or interrogation operations (e.g., interference checks) require the highest possible accuracy in the "digital model" used. This and the observation that *robust "noise removal" is, in general, impossible* imply our current strategy of scarcely smoothing images, and of thinning data only when this is implied by memory limitations of the computer. On the other hand, the "unit normal-vector image" $I_{nv} = \{\mathbf{n}_{ij}\}$, threshold estimation, and the regions-test are extensively used by the new RG and VOF procedures, as they lead to an improved surface-reconstruction algorithm.

# 3 Constructing Surfaces that Capture "Design Intent"

As the primary objective of this research is to serve the computerization of automobile styling, we state the surface-reconstruction problem as follows:

PROBLEM $\mathcal{S}$. **Given input:** the DRI $I$, a distance-error tolerance $T$, a normal-compatibility tolerance $V$ $(0 < V < 1)$, and a maximum permissible degree $t$ for bivariate polynomial approximations. **Decompose** $I$ into $N$ quasidisjoint *regions* (i.e., nonoverlapping subsets that may intersect only at their boundaries) $R_q$, $q = 1, 2, ..., N$, and define $N$ polynomial functions $z_q = z_q(x,y)$, $q = 1, 2, ..., N$, such that:

**(i)** For every point $\mathbf{p}$ in a particular region $R_q$

$$\text{the distance of } \mathbf{p} \text{ to the surface } z_q \text{ is smaller than } T, \text{ and} \tag{2}$$

$$\mathbf{n}_{ij} * \mathbf{N}_{z_q}[\mathbf{p}_{ij}] \;>\; V, \tag{3}$$

where $\mathbf{N}_{z_q}[\mathbf{p}_{ij}]$ denotes the positive normal vector of $z_q$ evaluated at the orthogonal projection of $\mathbf{p}_{ij}$ onto $z_q = z_q(x,y)$, and the symbol "$*$" denotes the vector inner-product operation.

**(ii)** The total number of regions (and surfaces) $N$ is as small as possible.

**(iii)** Each region $R_q$ should include as few interior boundaries (holes) as possible. (Ideally, each interior boundary of $R_q$ should correspond to a cavity or hole or sharp edge in the physical object.)

**(iv)** Each region $R_q$ should correspond to a *domain* $d(R_q)$ — the domain of any subset $R$ of $I$ is the smallest rectangle of the plane $z = 0$, whose sides are parallel to the $x$ and $y$ axes, and includes all points $(x_i, y_i, 0)$, where $(x_i, y_i, z_{ij}) \in R$ — that is as close as possible to a square. Also, for any two regions $R_q$, $R_l$ that intersect (at their boundaries) the projection of $R_q \cap R_l$ onto the plane $z = 0$ should be as close as possible to a straight line.
**(v)** Each surface $z_q$ should be defined by polynomials of low degree including the least possible degrees-of-freedom (d.o.f.s).
**(vi)** Each surface $z_q$ should be well-defined everywhere in the domain of the image $I$ (not only in the domain of the corresponding region $R_q$), and have a sufficiently smooth (: gradually varying) gaussian and mean curvature.

The above problem focuses on producing *extended fair surfaces* that reveal a stylist's intention, instead of a "complete surface-model" of patches meeting with a prescribed level of continuity [5][7][8]. Thus, the surface-fitting method used must allow regions (and surfaces) to extend beyond sharp edges and discontinuities in the data (see Fig. 2). Regarding the parameters involved in the description of Problem $\mathcal{S}$, we note that typically the maximum permissible degree for polynomials is 4×4, and, for high-quality DRIs of automobile components, one uses $T \approx 0.1 - 0.5\ mm$ and $V \approx 0.96 - 0.995$.

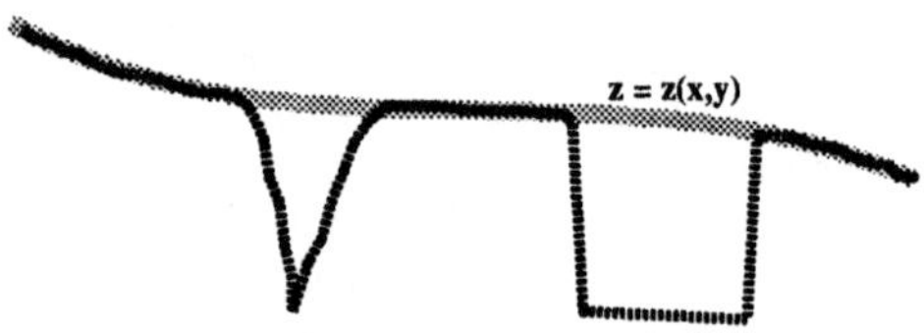


Fig. 2 2-D DRI (dotted line) with disconnected parts that can be fitted to the same surface (grey line).

## 4 Variable-Order Surface Fitting by Region Growing

In [1]—[4] a variety of algorithms based on RG and VOF have been described and applied to image segmentation/understanding problems with good results. The combined Variable-OrderFitting*by*RegionGrowing (VOFRG) algorithm, given in Table 1, simplifies many operations of the original procedures so that processing of very large data-sets is possible. Of

**Algorithm** Variable-OrderFitting*by*RegionGrowing (VOFRG)

| **Input**: $I, I_{nv}, t, T, V$; | **Output**: $\{R_q, z_q\}_{q=1}^N$; |
|---|---|

{ Symbols are defined in the description of Problem $\mathcal{S}$}

| **Working Variables/Symbols/Procedures**: $n$: number of regions, $d$: degree of current surface, $H$: threshold, *RegionGrows*: flag, *RemainingQuads*: integer, $D_q$: domain of current region, $NQ(R_q)$: number of points in $R_q$, $\not\gg$: "not significantly larger", $r(R_{q,m}, z_{q,m})$: regions-test procedure; |
|---|

```
n = 0; RemainingQuads = number of object quads in I;
while RemainingQuads > 0 begin {$}
  Increment n;  d = 1;   SelectSeedRegion(I, I_nv, R_n, D_n);
  while d ≤ t begin  {♮}
    LeastSq.PolynomialFitting(R_n, d, z_n);
    ThresholdSelection(R_n, z_n, H);   RegionGrows = true;
    while ( [RegionGrows == true] or [H < T] ) begin  {§}
      Local/GlobalRegionGrowing(R_n, D_n, z_n, d, H, V, T, RegionGrows);
    end  {§}
    if ( [NQ(R_n,d) ≯ NQ(R_n,d-1)] or [r(R_n,d-1, z_n,d-1) ≯ r(R_n,d, z_n,d)] )
      then  recreate R_n,d-1, z_n,d-1 and set d = t + 1;  else  d = d + 1;
  end  {♮}
  RemainingQuads = RemainingQuads - NQ(R_n);
end {$}
CreateFilesIGES144({z_q, R_q}_{q=1}^N);
End.
```

Tab. 1 Surface-Reconstruction Algorithm.

the many alternatives examined, the version of VOFRG reported here combines acceptable output (it produces surfaces that exactly satisfy requirement (i) of Problem $\mathcal{S}$ and approximately fulfil (ii)—(vi)) with sufficiently small execution time.

Growth of the region $R_{n,d}$ — wherever a second subscript appears in $R_n$ or $z_n$, this denotes the degree of the fitted polynomial — is initialized by specifying a seed-region $R_{n,d}^{k=0}$; [12] presents an efficient SelectSeedRegion algorithm, although users often prefer to control this step themselves. SelectSeedRegion is followed by application of VOF (see loop marked with {♮} in Table 1): First, a polynomial of degree $1 \times 1$ is fitted to the seed-region, followed by iterative application of Local/GlobalRegionGrowing (see below for a description), for polynomial degree $d = 1$, producing the region $R_{n,1}$ and the associated surface $z_{n,1}$. These are used as the seed-region (and starting surface) for reapplying Local/GlobalRegionGrowing with a polynomial degree $d = 2$, and so on. The result $\{R_{n,d}, z_{n,d}\}$ of the $d^{th}$ iteration, for $d = 2, 3, 4$, is compared with that of the previous iteration on the basis of two criteria: the size of the regions and the *regions-test*. If the result of the $d^{th}$ iteration fails to be superior to that of the previous iteration, with respect to *either one* of these criteria, the process is terminated and the final result is that of the iteration $d-1$, i.e., $\{R_{n,d-1}, z_{n,d-1}\}$. Of these two criteria, the first one is a by-product of region-growing while the second one requires some computational effort. The present implementation of the *regions-test* $r(R_{n,d}, z_{n,d})$ (see also discussion in §2) follows §4.7.2 of [1] and produces a quantity that approaches zero as the fit improves.

Regarding the *threshold* $H$ estimated by procedure ThresholdSelection (see also §2), this relates to the objective of growing the surface $z_n$ to closely approximate the underlying "physical" surface $z_f$. Thus, examination of compatibility of an image-point to $z_n$ (according to requirement (i) of Problem $\mathcal{S}$) should be based on an estimate of the *noise* in the DRI $I$ instead of the arbitrary limit $T$ [1]. Indeed, whenever the procedure Local/GlobalRegionGrowing examines the validity of the distance-error criterion (2), $T$ is replaced by the pre-calculated *threshold* $H$. Following [1], $H$ is calculated by multiplying the RMS fit error, between the seed-region and the corresponding surface, by a factor $W$ depending on the quality of $I$. The tests reported in [1] suggest $W = 2.5$ for good-quality images, which is the value used also here.

Completion of the main part (loop marked with {$}) of the VOFRG Algorithm is followed by application of the procedure CreateFilesIGES144 that converts each polynomial $z_q$ to the IGES 144 [11] format, i.e., to a parametric B-spline trimmed in accordance to $R_q$. This may be preceded by a procedure [12] that reduces position- and tangent-plane-discontinuities between adjacent polynomial surfaces using a variant of the procedure `GrowRegionOUT` described below.

## 4.1 The Procedure Local/GlobalRegionGrowing

This procedure takes as primary input, for creating the $n^{th}$ region, a seed-region $R_n^{k=0}$, a fitted surface $z_n^{k=0}$, and a threshold $H$. Growth is attempted first within $d(R_n^{k=0})$ and then outside this domain considering only points from $I$ that do not belong to the already completed regions $R_1, R_2, \ldots, R_{n-1}$. Region growing is an iterative process, where first points that are compatible with the current surface $z_n^k$ are found and collected in the new region $R_n^{k+1}$, and then a new surface $z_n^{k+1}$ is fitted to $R_n^{k+1}$, in the least-squares sense. The procedure, which is a simplified version of that in [1], is detailed in the paper [11]; below only a short overview is given.

(A) LocalRegionGrowing.

(A.1) `GrowRegionIN` *(growing a region within its domain):* Here, only points $\mathbf{p}_{ij}$ with $(i, j)$ in $d(R_n^k)$ are considered. Those satisfying (2) — with $T$ replaced by the current threshold $H$ — and (3) are collected in $R_n^{k+1}$ and a new surface $z_n^{k+1}$ is calculated by least-squares fitting. This process is iterated until a sufficiently large percentage of the points corresponding to $d(R_n)$ are in $R_n$, or $R_n$ has stopped growing.

(A.2) `GrowRegionOUT` *(growing a region outside of its domain):* This routine grows $R_n$ outside $d(R_n)$, towards "left" (: smaller column numbers), "right" (: larger column numbers), "down" (: smaller row numbers), and "up" (: larger row numbers); see Fig. 1. Rules are employed that specify appropriate growth directions so that the region grows uniformly towards the four possible directions and does not create "gaps" (by skipping quads) as implied by requirements (iii) and (iv) of Problem $\mathcal{S}$. More specifically, `GrowRegionOUT` always tries to grow $R_n$ in a small number $N_{gr}$ of columns/rows adjacent to the domain $d(R_n)$. The current implementation uses $N_{gr} = min\{K, L\}/50$, where $K$ and $L$ are the number of columns and rows in $I$, respectively.

Completion of the above procedures is followed by a comparison of the number of points in $R_n$ to that before applying local region-growing, for evaluating progress. When this is found unsatisfactory, the threshold $H$ is increased and local region-growing is reapplied to $R_n$. Local Region-growing is terminated when $R_n$ has stopped growing or $H$ has reached the user-specified limit $T$.

(B) GlobalRegionGrowing.
In accordance to requirement (ii) of Problem $\mathcal{S}$ (see §3), it is desirable that one identifies disconnected parts of $I$ that can be approximated by the same polynomial surface. This can be achieved by re-applying the procedure GrowRegionOUT, this time with a large $N_{gr}$, e.g., $N_{gr} = min\{K, L\}$, ensuring that the current region is allowed to grow everywhere on $I$.

## 5 Region-Growing with Implicit Surfaces

Polynomial representations, although very popular, do not cover all industry's requirements. Sometimes other representations are preferable, e.g., the implicit form of quadrics for images extracted from quadric physical surfaces. The ideas of VOF and RG can also be applied to these surfaces with only two prerequisites: A procedure to fit an implicit quadric to a point-set and a procedure to project a point onto a quadric surface. The former is readily available although least-squares fitting of quadric surfaces is not as reliable as its polynomial counterpart (see [6], [9], [11] and comments below). Projecting a point onto a quadric can be accomplished using a Newton-Raphson-type iterative procedure with initial estimates of projective direction and distance obtained using the formulas proposed in [14].

The present version of the Quadric Region-Growing (QRG) algorithm differs from the one for polynomials in that a much "less-aggressive" strategy is used in increasing the threshold $H$ which controls local region-growing. This causes a much slower growth of a region but has succeeded in eliminating failures of the technique occurring for quadrics with "extra branches" [11]. Still, there are cases where QRG fails to converge, e.g., for a point-set lying closely to an almost flat ellipsoid (see Fig. 3), or to a cylinder with a very large radius. *Our current work* focuses exactly on these problematic cases. One of the approaches investigated is the development of an improved quadric-fitting routine according to the ideas described in [6].

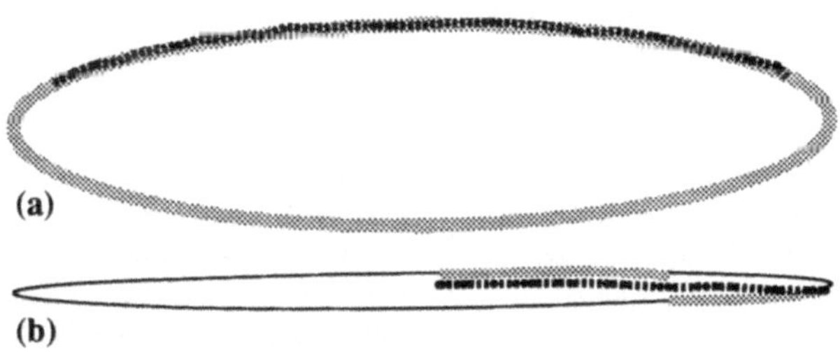


Fig. 3 **(a)** Quadric region-growing produces an acceptable surface. **(b)** Quadric region-growing associates this region (dotted line) to a disconnected surface (grey line).

# 6 Conclusions

Both the polynomial and the implicit-quadric versions of VOFRG have been applied to numerous images, produced at GM, with satisfactory results regarding the curvature of the final surfaces and the positional- and tangent-plane (dis)continuity between them. Regarding the total number of surfaces and the number of d.o.f.s in each one of them, the obtained results are superior to those produced by standard GM practices, which typically use a B-spline surface with 12-20 segments of degree $3 \times 3$ to represent point-data approximated by a single VOFRG polynomial of degree up to $4 \times 4$.

## 6.1 Stimulus Bound Image Analysis

Variable-order surface fitting may be thought of as a hypothesis testing algorithm where *the input data* can automatically change a selection of surface-model (hypothesis), thus the final result is bound by the given range-image. In Image Processing this short of techniques are characterized as *"stimulus bound"* where the stimulus is the original sensed image (see, e.g., §4.2.3 in [1]).

In a stimulus-bound process, all operations (sub-processes) are *bound* to the original stimulus to reduce the probability of data interpretation errors. A typical example of a process which is not stimulus-bound is the standard *Points* $\longrightarrow$ *Curves* $\longrightarrow$ *Surfaces (PCS)* methodology for creating a surface-model. PCS is using, in general, the original image merely for constructing a network of curves which is the only input used in the surface creation stage. The authors hope that the results produced by VOFRG will convince CAD researchers to consider stimulus-bound techniques (and *simple-to-complex hypothesis testing* [1]) in surface-reconstruction, especially for high-quality images that require a very small "point-to-surface" error.

**Acknowledgements:** The work of N. Sapidis was supported by the GM Design Center, the GM R&D Center, and the Ship-Design Lab of NTUA. We thank J. Stevens, T. Sanderson, R. Sarraga, R. Tilove, D. Field, W. Frey, and D. Papageorgiou for data (images) and many useful comments during all stages of this research. Also, personal communications with G. Golub on least-squares fitting of quadrics are gratefully acknowledged.

Nickolas S. Sapidis
*National Technical University of Athens*
Department of Naval Architecture &
Marine Engineering, c/o P. Kaklis
9, Heroon Polytechneiou Str.
Zografou, Athens 157-73, GREECE
*sapidis@deslab.ntua.gr*

Paul J. Besl
*General Motors Corporation*
Research & Development Center
Manufacturing and Design Systems
30500 Mound Road, Box 9055
Warren, MI 48090-9055, USA
*besl@predator.cs.gmr.com*

# References

[1] Besl, P.J.: *Surfaces in Range Image Understanding.* Springer-Verlag, New York, 1988.

[2] Besl, P.J.: *Geometric Signal Processing.* In: Jain, R., Jain, A. (eds), *Range Image Understanding*, Springer-Verlag, 1989.

[3] Besl, P.J.: *Active Optical Range Imaging Sensors.* In: Sanz, J. (ed), Advances in Machine Vision: Architectures and Applications, Springer-Verlag, New York, 1989. (See also: Machine Vision and Applications **1**, 127-152.)

[4] Besl, P.J., Jain, R.C.: *Segmentation Through Variable-Order Surface Fitting.* IEEE Transactions on Pattern Analysis and Machine Intelligence **10** (1988), 167-192.

[5] Dietz, U.: *Erzeugung Glatter Flächen aus Meßpunkten.* Preprint-Nr. 1717, FB Mathematik, TH Darmstadt, Febr. 1995.

[6] Gander, W., Golub, G.H., Strebel, R.: *Fitting of Circles and Ellipses–Least Squares Solution.* Manuscript SCCM-94-08, Scientific Computing and Computational Mathematics Program, Stanford University, Aug. 1994.

[7] Gu, P., Yan, X.: *Neural Network Approach to the Reconstruction of Freeform Surfaces for Reverse Engineering.* Computer-Aided Design **27** (1995), 59-64.

[8] Milroy, M.J, Bradley, C., Vickers, G.W., Weir, D.J.: *$G^1$ Continuity of B-spline Surface Patches in Reverse Engineering.* Computer-Aided Design **27** (1995), 471-478.

[9] Moore, D., Warren, J.: *Approximation of Dense Scattered Data Using Algebraic Surfaces.* Technical Report *TR90-135*, Department of Computer Science, Rice University, 1990.

[10] Sapidis, N. (ed): *Designing Fair Curves and Surfaces: Shape Quality in Geometric Modeling and CAD.* Society for Industrial and Applied Mathematics (SIAM), Philadelphia, 1994.

[11] Sapidis, N., Besl, P.J.: *Direct Construction of Polynomial Surfaces from Dense Range Images through Region Growing.* ACM: Transactions on Graphics **14** (1995), 171-200.

[12] Sapidis, N.: *Variable-Order Surface Reconstruction Through Region Growing.* To appear in: Roller, D. and Brunet, P. (eds), CAD Tools for Products, Springer-Verlag.

[13] Sarkar, B., Menq, C.-H.: *Smooth Surface Approximation and Reverse Engineering.* Computer-Aided Design **23** (1991), 623-628.

[14] Taubin, G.: *Algebraic Nonplanar Curve and Surface Estimation in 3-Space with Applications to Position Estimation.* Technical Report LEMS-43, Division of Engineering, Brown University, Providence, 1988.

# Smooth B–Spline Surface Approximation to Scattered Data

*Josef Hoschek, Ulrich Dietz*

Technische Hochschule Darmstadt
E–mail: hoschek@mathematik.th–darmstadt.de
udietz@mathematik.th-darmstadt.de

**Abstract:** An algorithm for approximation of arbitrary clouds of points with integral tensor product B–spline surfaces is presented. The clouds may be scattered may have holes and may have arbitrary boundaries. The usual methods in Reverse Engineering subdivide the given cloud into rectangular parts and approximate these parts individually. In the presented paper an overall algorithm for tensor product B–spline approximation with free boundary curves is introduced.

## 1 Introduction

If the boundaries of the given cloud of points have a rectangular structure the following **strategy** can be used [7].

1. Approximate the boundary points by four boundary curves (B–spline curves) with parameter correction (s. section 4).
2. Construct within these four boundary curves a bilinear Coons–patch $C$ and project (perpendicular to $C$) the interior points $\mathbf{P}_i (i = 1, ..., r)$ of the given cloud on $C$. In the corresponding parameter domain of $C$ an initial guess of the parameter values $(u_i, v_i)$ of $\mathbf{P}_i$ can be calculated.
3. With this initial guess of the parametrization of the points $\mathbf{P}_i$ the given set of points is approximated by one tensor product B–spline surface of fixed order $k$ in the parametric representation (s. [2, 7])

$$\mathbf{X}(u, v) = \sum_{j=0}^{n} \sum_{l=0}^{m} \mathbf{d}_{jl} N_{jk}(u) N_{lk}(v) \tag{1}$$

where the parameters $u, v$ are defined over suitable chosen knot vectors while the $\mathbf{d}_{jl}$ are the unknown control points (de Boor points). The objective function of the approximation problem is

$$Q = \sum_{i=1}^{r} (\mathbf{P}_i - \sum_{j=0}^{n} \sum_{l=0}^{m} \mathbf{d}_{jl} N_{jk}(u_i) N_{lk}(v_i))^2 \to \min. \tag{2}$$

The solution of (2) is obtained by wellknown least square techniques.

4. This approximation step leads to an approximation surface $\mathbf{X}$. To get an improved parametrization of the given points we replace the parameter values for each point by those of the closest point on the surface (s. section 4). Then we restart the approximation process according to step 3.

Steps 3 and 4 are repeated until the maximal value of the distance error

$$\max_{i=1,\ldots,r} \mid \mathbf{P}_i - \mathbf{X}(u_i, v_i) \mid < \varepsilon. \tag{3}$$

Although the parameter correction is a very powerful tool, sometimes the approximation fails and unwanted oscillations appear (s. Fig. 1).

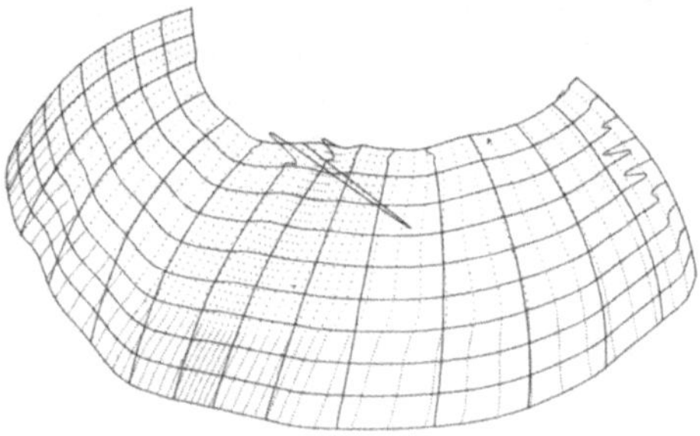

Fig. 1 Example where the least squares fit fails

There may be several reasons for undesired results: the given points have different densities (or even holes [14]), the polynomial degree is very high, the approximation is near interpolation (less points with respect to the given number of control points).

## 2 Approximation with energy terms

We obtain an extension of this approximation method if we exchange the total error sum (2) by a new error functional

$$\tilde{Q} = Q + \lambda \sum_p \alpha_p Q_p \rightarrow \min \tag{4}$$

where $Q_p$ are energy terms. $\lambda \geq 0$ is a penalty factor and the $\alpha_p$ are factors of a convex combination with $\sum \alpha_p = 1$. These factors $\alpha_p$ are called design parameters which allow the user to control the influence of the individual energies. Whereas, the penalty factor $\lambda$ (which will be calculated automatically) determines the influence of all the energy terms compared with the total error sum.

To get a linear system for the solution of (4), the energy terms $Q_p$ must be quadratic in the unknown control points. Often used energy functionals [1, 5, 15, 16] are

$$Q_1 = \iint (\text{grad}\, \mathbf{X})^2 \, du\, dv = \iint (\mathbf{X}_u^2 + \mathbf{X}_v^2) \, du\, dv \tag{5}$$

$$Q_2 = \iint (\mathbf{X}_{uu} + \mathbf{X}_{vv})^2 - 2(1-\mu)(\mathbf{X}_{uu}\mathbf{X}_{vv} - \mathbf{X}_{uv}^2)\, du\, dv \tag{6}$$

$$\begin{aligned} Q_3 &= \iint (\text{grad div grad } \mathbf{X})^2\, du\, dv \\ &= \iint (\mathbf{X}_{uuu} + \mathbf{X}_{uvv})^2 + (\mathbf{X}_{uuv} + \mathbf{X}_{vvv})^2\, du\, dv \end{aligned} \tag{7}$$

It is obvious that all these energies are quadratic in the control points. The energy $Q_1$ is a simple approximation of the area of surface $\mathbf{X}$. The second energy is a small deflection approximation for the exact bending energy of a thin plate. If we set the so called Poisson ratio $\mu$ to zero we obtain an approximation for the integral $\iint \kappa_1^2 + \kappa_2^2\, du\, dv$ with the principal curvatures $\kappa_1$, $\kappa_2$. Finally, the functional $Q_3$ is a measure for the change of curvature of surface $\mathbf{X}$. Other energy or fairness functionals and their application to surface modeling can be found in [6, 11, 13, 17].

In contrast to the energy functionals proposed by Greiner [3] the above functionals depend on the actual parametrization of surface $\mathbf{X}$. Since the functionals of Greiner are also quadratic in the control points and in general yield better approximations to the 'real' energies they could be used instead of the $Q_i$'s. As the computational evaluation of Greiner's functionals is much more expensive we do use here the functionals $Q_i$.

Differentiation of (4) leads to the necessary conditions

$$\frac{\partial \tilde{Q}}{\partial \mathbf{d}_{jl}} = \frac{\partial Q}{\partial \mathbf{d}_{jl}} + \lambda \sum_p \alpha_p \frac{\partial Q_p}{\partial \mathbf{d}_{jl}} \stackrel{!}{=} 0 \tag{8}$$

or the linear system

$$(A + \lambda \sum_p \alpha_p A_p)\, d = b \qquad \Leftrightarrow \qquad \tilde{A}\, d = b \tag{9}$$

The elements of the matrices $A_p$ are given by

$$\begin{aligned} A_1 :&\quad U_{jg}^{11} V_{lh}^{00} + U_{jg}^{00} V_{lh}^{11} \\ A_2 :&\quad U_{jg}^{22} V_{lh}^{00} + \mu U_{jg}^{20} V_{lh}^{02} + \mu U_{jg}^{02} V_{lh}^{20} + U_{jg}^{00} V_{lh}^{22} + 2(1-\mu) U_{jg}^{11} V_{lh}^{11} \\ A_3 :&\quad U_{jg}^{33} V_{lh}^{00} + U_{jg}^{31} V_{lh}^{02} + U_{jg}^{13} V_{lh}^{20} + U_{jg}^{11} V_{lh}^{22} \\ &\quad + U_{jg}^{22} V_{lh}^{11} + U_{jg}^{02} V_{lh}^{31} + U_{jg}^{20} V_{lh}^{13} + U_{jg}^{00} V_{lh}^{33} \end{aligned} \tag{10}$$

where

$$\begin{aligned} U_{jg}^{st} &= \int N_{jk}^{(s)}(u) N_{gk}^{(t)}(u)\, du, \\ V_{lh}^{st} &= \int N_{lk}^{(s)}(v) N_{hk}^{(t)}(v)\, dv. \end{aligned} \tag{11}$$

The occuring integrals can be calculated either directly or numerically exact by Gaussian quadrature. The highest involved derivatives for functional $Q_p$ are the derivatives of order $p$.

The vector $d$ contains the unknown control points. $A$ is the square matrix of the usual normal equation system and $b$ the corresponding right hand side. The square matrices $A, A_p, \tilde{A}$ are symmetric, positive semi-definite and have a banded structure because of the locality of the B-spline basis functions.

If there is at least one $\alpha_p > 0$ and the points are in a general situation the matrix $\tilde{A}$ is regular and the solution of the system is uniquely determined. In other words we obtain a well determined approximation surface $\mathbf{X}$ for arbitrary point distributions even if there are holes in the point cloud or if the point cloud is very sparse.

# 3 Free boundaries and parametrization

In the first section we made special assumptions concerning the given point cloud. Therefore, the boundaries of the points could directly be used to calculate the boundaries of the reference surface and also of the tensor product approximation surface.

Now we make no assumptions concerning the boundaries of the points; they may be arbitrary and must not be given explicitly. Generally the boundaries of the point set will lie totally in the interior of the tensor product approximation surface. In a last step the unnecessary surface parts are trimmed away by curves defined in the parametric domain of the surface (s. section 5).

The initial parametrization of the point set is calculated with help of a reference surface. The only condition which must be be fulfilled by the reference surface is that the perpendicular projection of the points onto the surface is bijective. For many data sets the least squares fitting plane will be sufficient. Sophisticated reference surfaces as proposed in [7, 10] could be used if the structure of the point set is more complicated.

The initial parameter values $(u_i, v_i)$ are calculated as the parameter values of the perpendicular projections of the points $\mathbf{P}_i$ onto the reference surface. Since these parameter values are only a first guess and are optimized in the subsequent steps the quality of this first parametrization is not important. Therefore, the simplest parametrization compatible with the point cloud should be used, e.g. the projection onto a plane.

# 4 Parameter correction

With the determined initial parametrization for the point cloud an approximation surface $\mathbf{X}$ can be calculated by minimizing the quadratic objective functional $Q$ resp. $\tilde{Q}$. But the obtained surface $\mathbf{X}$ is not minimizing the shortest distances but some arbitrary distances induced by the parametrization.

The idea of the parameter correction is to improve the parametrization with help of the surface $\mathbf{X}$ and then to redo the fitting step [4, 8, 12]. Both steps of reparametrization and approximation are repeated in a loop until an end condition is satisfied (e.g. approximation error is small enough).

For each point $\mathbf{P}_i$ the corresponding parameter values $(u_i, v_i)$ are replaced by the parameter values $(\tilde{u_i}, \tilde{v_i})$ of the closest point to $\mathbf{P}_i$ on the surface $\mathbf{X}$. This is done by a Newton method which is local convergent of order 2. The correction terms are given by

$$\begin{pmatrix} \Delta u \\ \Delta v \end{pmatrix}^{(k)} = \begin{pmatrix} \mathbf{X}_u^2 + (\mathbf{X}-\mathbf{P})\mathbf{X}_{uu} & \mathbf{X}_u\mathbf{X}_v + (\mathbf{X}-\mathbf{P})\mathbf{X}_{uv} \\ \mathbf{X}_u\mathbf{X}_v + (\mathbf{X}-\mathbf{P})\mathbf{X}_{uv} & \mathbf{X}_v^2 + (\mathbf{X}-\mathbf{P})\mathbf{X}_{vv} \end{pmatrix}^{-1} \begin{pmatrix} (\mathbf{X}-\mathbf{P})\mathbf{X}_u \\ (\mathbf{X}-\mathbf{P})\mathbf{X}_v \end{pmatrix} \quad (12)$$

Generally one obtains a sufficient accuracy after only 1–3 iterations. In combination with the simple least squares fit parameter correction may not be stable depending on the point distribution. Oscillations may occur in the surface which will increase during the iterations. The parameter correction together with the approximation including energy terms is very robust. The energy terms ensure that the approximation surfaces are smooth and stiff and so the parameter correction can work properly without stopping too early in any local minima.

## 5 Trimming

For the final approximation surface the boundary curves will in general not coincide with the boundaries of the point cloud. So the surface regions outside the point set should be trimmed away by B–spline curves defined in the parametric domain of the surface.

The boundary points $\mathbf{R}_i$ are projected perpendicular onto the surface and then for each projected point the preimage $(u_i, v_i)$ in the parametric domain is calculated.

These points $(u_i, v_i)$ are approximated by a 2d B–spline curve [9]

$$\mathbf{C}(t) = (u(t), v(t)) = \sum_i \mathbf{d}_i N_{ik}(t) \quad (13)$$

with least squares techniques and parameter correction. The final trimming curve is obtained by inserting $\mathbf{C}(t)$ into the surface representation of $\mathbf{X}$, i.e. $\mathbf{X}(u(t), v(t))$. In a similar way holes can be performed in the interior of the surface if the point cloud requires it.

## 6 The Algorithm

The tools described in the previous sections are now combined into an automatic algorithm for smooth approximation of point sets with free boundaries.

1. Set a suitable order $k$, knot vectors $U, V$, numbers of control points $n, m$, an error tolerance $\varepsilon_0$ and the penalty factor $\lambda = \lambda_0 \gg 1$.
2. Calculate an initial guess (a rough approximation) for the parameter values $(u_i, v_i)$.
3. Solve the linear approximation problem (9).
4. Correct the parameter values $(u_i, v_i)$.
5. Calculate the maximal point distance $\varepsilon$ to the surface.
6. Decrease $\lambda$ (e.g. $\lambda := \lambda/2$). Proceed with step 3 if $\varepsilon > \varepsilon_0$ and the maximal iteration number isn't reached.
7. Cut off the unnecessary surface parts outside the point cloud.

At the beginning the algorithm starts with a parametrization which is in general far away from the final (optimal) one. For the first calculated surfaces the objective functional is dominated by the energy terms. Therefore, the surfaces are stiff and of simple shape allowing the parameter correction to improve the parameters in a stable way. In every iteration step the penalty factor $\lambda$ is lowered. This allows the surface to move towards the point cloud step by step. The result of the algorithm is the smoothest surface satisfying the given error tolerance. This error tolerance must be chosen suitable in respect to the point set. If there is a disturbed point cloud and very small error bound the resulting approximation surface will be disturbed as well.

# 7 Benchmarks

In this section we present the results of the described algorithm for the 6 provided benchmark examples. The initial parametrization for all examples except the last one is obtained by the projection of the points to the least squares fitting plane. To get the high accuracy in the last example chordal parametrization and manual knot placement near the sharp edges has been used. The thin grey lines in the figures represent the patch boundaries. The *isophotes* which should demonstrate the quality of the approximation results are drawn as thick black lines.

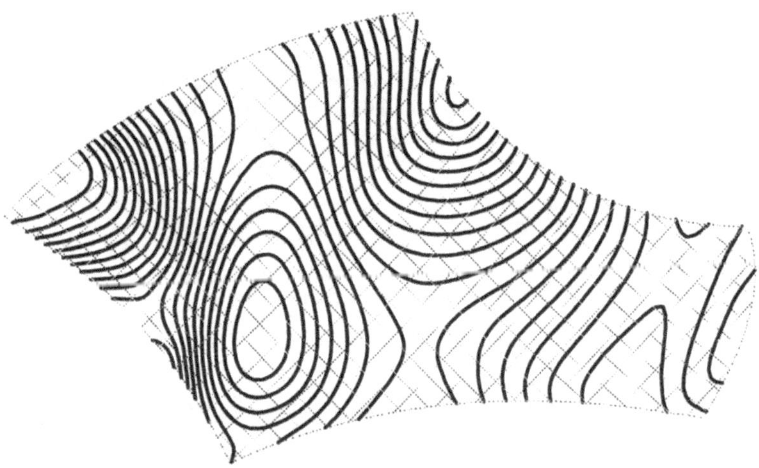

Fig. 2 benchmark 1:

- 842 points on a five-sided area; diameter $D = 179.5$
- trimmed surface with $25 \times 25$ bicubic patches minimizing thin plate energy; 20 iterations
- maximum deviation $\delta_{max} = 0.3215 \widehat{=} 0.18\ \%\ D$
- averaged deviation $\delta_{avg} = 0.0477 \widehat{=} 0.026\ \%\ D$

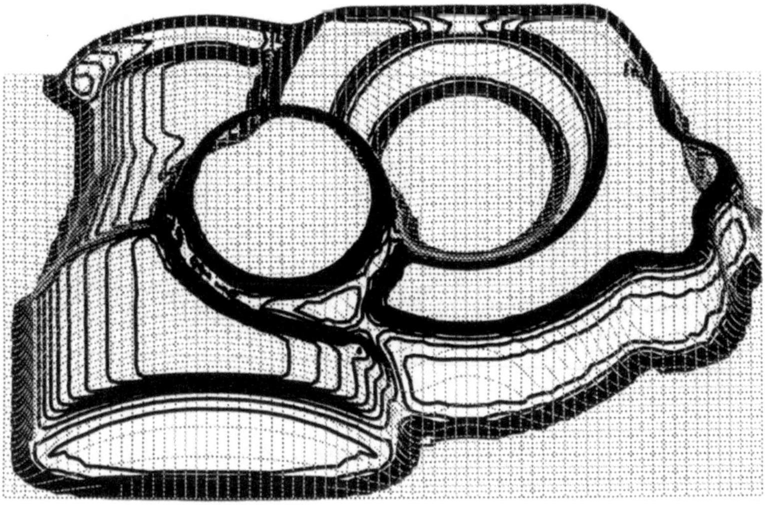

Fig. 3 benchmark 2:

- 4101 points scanned from top of a metal part; diameter $D = 98.7$
- untrimmed surface with 70 × 70 bicubic patches minimizing '0.1 surface area + 0.9 thin plate energy'; fixed boundaries; 20 iterations
- maximum deviation $\delta_{max} = 0.630 \mathrel{\widehat{=}} 0.64\ \%\ D$
- averaged deviation $\delta_{avg} = 0.027 \mathrel{\widehat{=}} 0.027\ \%\ D$

# 8 Concluding remarks

The presented algorithm yields smooth surfaces to a prescribed error tolerance with only a view variables of user input. The algorithm can be generalized in several directions. The global penalty factor $\lambda$ for the energy terms could be made local. An improvement in the reconstruction of sharp edges and areas of large curvature is expected. Additionally the approximation of point normals $\mathbf{N}_i$ could be included in the algorithm. The approximation step will remain linear if the objective functional is extended by the term $Q_N = \sum_i \mathbf{N}_i \mathbf{X}_u + \sum_i \mathbf{N}_i \mathbf{X}_v$. Probably the quality of the final surface could be improved if one uses the functionals of Greiner in the last iteration step instead of the functionals $Q_p$. Also we can consider the overdetermined system of residuals instead of the normal equations. It can be solved very stable with orthogonalisation methods. The mathematics of the algorithm are only outlined in this paper. A more detailed description will be given in the proceedings of the *Advanced Course on Fairshape / Lambrecht 1996*.

# 9 Acknowledgment

The research work of the second author is supported by the german *Bundesministerium für Bildung, Wissenschaft, Forschung und Technologie (BMBF)*.

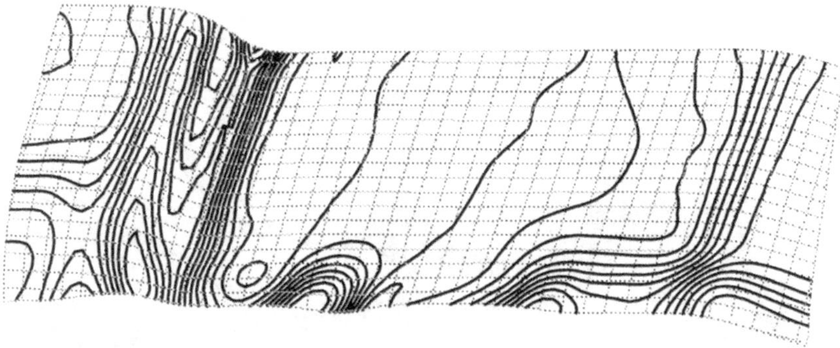

Fig. 4 benchmark 3:

- 1827 points from a flat object with extremly varying density; diameter $D = 381.0$
- trimmed surface with 20 × 40 bicubic patches minimizing thin plate energy; 30 iterations
- maximum deviation $\delta_{max} = 0.1869 \widehat{=} 0.049$ % $D$
- averaged deviation $\delta_{avg} = 0.0173 \widehat{=} 0.0045$ % $D$

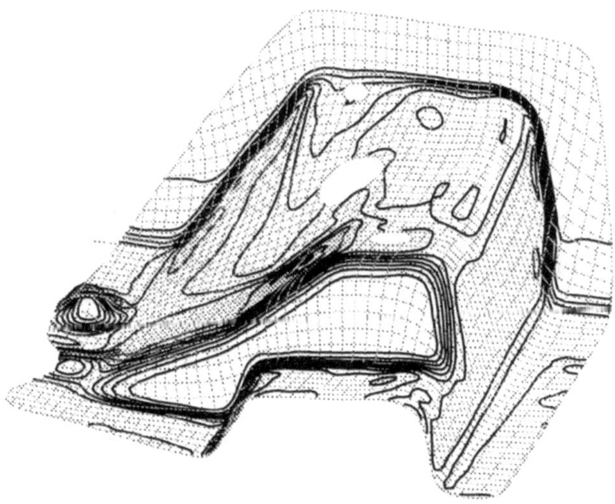

Fig. 5 benchmark 4:

- 95986 points; dense scan with holes; diameter $D = 223.5$
- trimmed surface with 50 × 50 bicubic patches minimizing thin plate energy; 12 iterations
- maximum deviation $\delta_{max} = 2.732 \widehat{=} 1.2$ % $D$
- averaged deviation $\delta_{avg} = 0.045 \widehat{=} 0.02$ % $D$

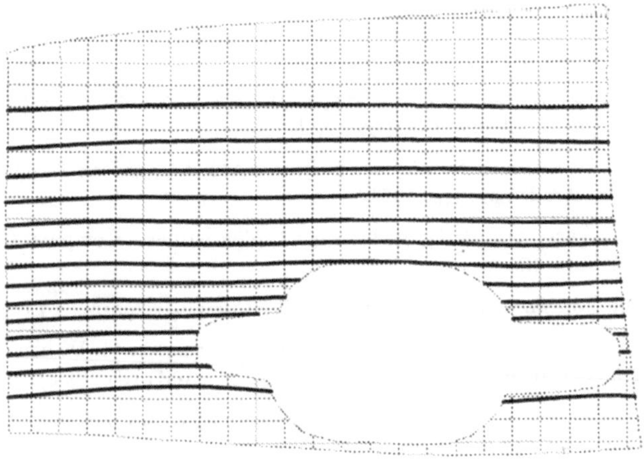

Fig. 6 benchmark 5:

- 105709 points; very dense scan of a car door; diameter $D = 475.6$
- trimmed surface with $25 \times 25$ bicubic patches minimizing thin plate energy; 24 iterations
- maximum deviation $\delta_{max} = 0.8026 \widehat{=} 0.16\ \%\ D$
- averaged deviation $\delta_{avg} = 0.056 \widehat{=} 0.01\ \%\ D$

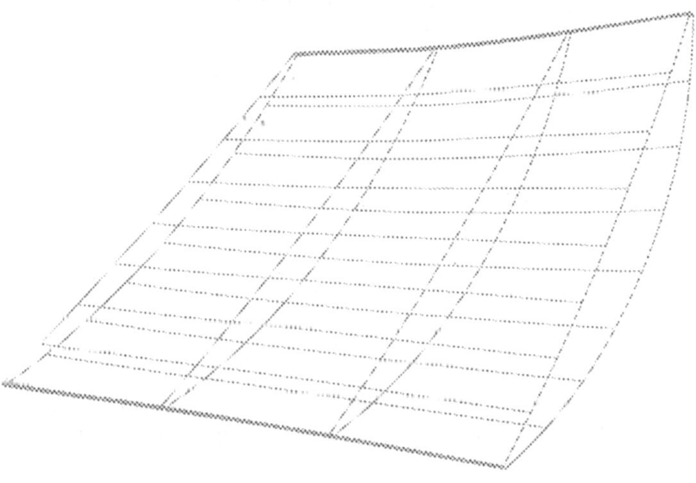

Fig. 7 benchmark 6:

- 1546 points in plane sections from a turbine blade; diameter $D = 61.3$
- untrimmed surface with $28 \times 3$ bicubic patches; no energy; manual knot placement; no iterations
- maximum deviation $\delta_{max} = 0.0008 \widehat{=} 0.0013\ \%\ D$
- averaged deviation $\delta_{avg} = 0.0002 \widehat{=} 0.00032\ \%\ D$

# References


[1] Bloor, M. I. G., Wilson, M. J. and Hagen, H.: *The smoothing properties of variational schemes for surface design.* Computer Aided Geometric Design **12** (1995), 381–394.

[2] Farin, G.: *Curves and Surfaces for Computer Aided Geometric Design – A Practical Guide.* 3. ed. Academic Press 1993.

[3] Greiner, G.: *Variational design and fairing of spline surfaces.* Computer Graphics Forum **13**:3 (1994), 143–154.

[4] Grossmann, M.: *Parametric curve fitting.* The Computer Journal **14** (1970), 169–172.

[5] Hadenfeld, J.: *Local energy fairing of B–spline surfaces.* In M. Dæhlen, T. Lyche, and L. L. Schumaker (eds.): Mathematical Methods in CAGD III, (1995), 203–212.

[6] Hagen, H., Schulze, G.: *Automatic smoothing with geometric surface patches.* Computer Aided Geometric Design **4** (1987), 231–235.

[7] Hoschek, J., Lasser, D.: *Grundlagen der geometrischen Datenverarbeitung.* 2. Auflage, Teubner 1992.

[8] Hoschek, J., Schneider, F.-J., Wassum, P.: *Optimal approximate conversion of spline surfaces.* Computer Aided Geometric Design **6** (1989), 293–306.

[9] Hoschek, J., Schneider, F.-J.: *Approximate spline conversion for integral and rational Bézier and B-spline surfaces.* In R. E. Barnhill (ed.): Geometry Processing for Design and Manufacturing, SIAM (1992), 45–86.

[10] Ma, W., Kruth, J. P.: *Parametrization of randomly measured points for least squares fitting of B–Spline curves and surfaces.* Computer–Aided Design **27** (1995), 663–675.

[11] Moreton, H. P., Séquin, C. H.: *Functional optimization for fair surface design.* ACM Computer Graphics **26** (1992), 167–176.

[12] Pratt, M. J.: *Smooth parametric surface approximations to discrete data.* Computer Aided Geometric Design **2** (1985), 165–171.

[13] Rando, T., Roulier, J.A.: *Designing faired parametric surfaces.* Computer–Aided Design **23** (1991), 492–497.

[14] Schmidt, R. M.: *Fitting scattered surface data with large gaps.* Surfaces in CAGD, R. E. Barnhill, W. Boehm (eds.), (1983), 185–189.

[15] Sinha, S. S., Schunck, B. G.: *A two-stage algorithm for discontinuity-preserving surface reconstruction.* IEEE Transactions on Pattern Analysis and Machine Intelligence **14** (1992), 36–55.

[16] Terzopoulos, D.: *Regularization of inverse visual problems involving discontinuities.* IEEE Transactions on Pattern Analysis and Machine Intelligence **8** (1986), 413–424.

[17] Welch, W., Witkin, A.: *Variational surface modeling.* ACM Computer Graphics **26** (1992), 157–166.

# Reverse Engineering using B-splines

*Matthias Eck*[1], *Hugues Hoppe*[2]

[1] University of Darmstadt, Germany
[2] Microsoft Corporation, USA

**Abstract:**
This paper briefly reviews a recent algorithm of the authors for reconstructing a network of tensor product B-spline surface from a set of unorganized 3D points. Unlike previous work which considers primarily the problem of fitting a single B-spline patch, this new procedure allows the reconstruction of surfaces of arbitrary topological type satisfying tangent plane continuity along common boundary curves. Furthermore, adaptive refinement of the patch network is performed to fulfill user-specified error tolerances.

## 1 Introduction

*Reverse engineering* refers to the reconstruction of geometric models from existing physical objects (Figure 1). Devices such as coordinate measuring machines and laser scanners are used to measure large collections of points on objects of interest. In Figure 2 such a scenario is simulated by randomly sampling 4000 points on the surface of Figure 1. The problem of converting these measurements into useful models is called *surface reconstruction*.

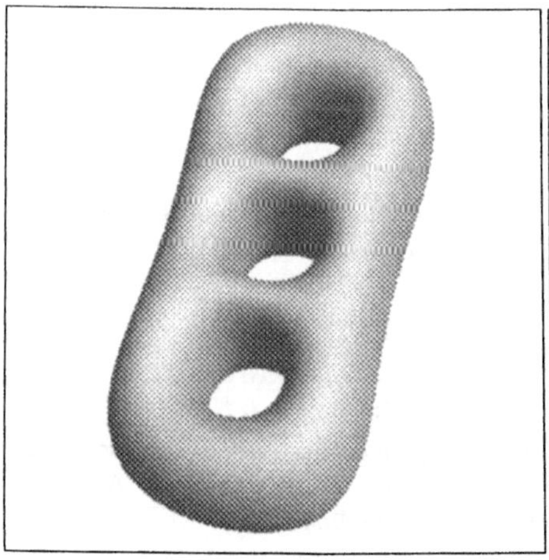

Fig. 1 An unknown surface $S_u$

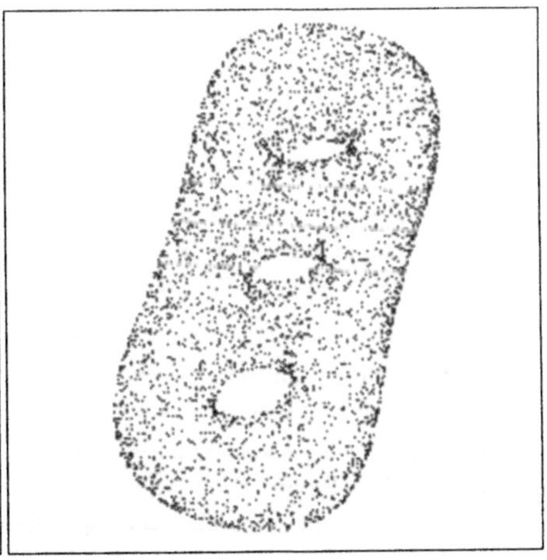

Fig. 2 A "measurement" of 4000 scattered points $P$

On the other hand, even with advanced geometric modeling systems, creating organic shapes such as human faces and free-form surfaces is still a challenging task. In most modeling systems, for better or worse, the ubiquitous smooth surface primitive is the four-sided tensor product B-spline patch. The general class of non-uniform rational B-splines (NURBS) is considered by many the de facto CAD standard.

Thus it was an open problem for a long time whether it would be possible to find a procedure for *automatically* reconstructing a B-spline surface $S$ of arbitrary topological type from an unorganized set of points $P = \{\mathbf{p}_1, \ldots, \mathbf{p}_N\}$. Here, two major difficulties occur:

1. Since a single B-spline patch can only represent surfaces of simple topological type (deformed planar regions, cylinders, and tori), a surface of arbitrary topological type must be defined as a network of B-spline patches. Automatically constructing both a network of patches and a parametrization of the data points over these patches is a difficult task.

2. The reconstructed B-spline patch network is often expected to be smooth (by which mostly tangent plane continuity or $G^1$ is meant). Enforcing $G^1$ continuity between adjacent patches while at the same time fitting the patch network to the points is a challenging problem.

In a recent paper of Eck and Hoppe [3] a solution to this very complex task was given. The purpose of the present paper is to briefly review the main steps of this new algorithm. Each step will thereby be illustrated with help of the dataset in Figure 2. The technical details, more examples and some remarks on previous work can be found in the original paper [3].

## 2 Algorithm

The B-spline surface reconstruction algorithm in [3] mainly consists of 5 successive steps which can be sketched as follows:

1. *Constructing an initial parametrization over a dense approximating mesh $M_0$:*

Initially, only a dense, unorganized set of points $P = \{\mathbf{p}_1, \ldots, \mathbf{p}_N\}$ is given, assumed to lie near some unknown surface $S_u$ (Figures 1 and 2). Using the surface reconstruction work of Hoppe et al. [4, 5], Step 1 constructs from such an unorganized set of points $P$ an initial surface approximation in the form of a dense triangular mesh $M_0$ (Figure 3). Then the points $P$ are projected onto $M_0$ to obtain their initial parametrizations.

The purpose in constructing $M_0$ is basically to find a parametric domain of the same topological type as the unknown surface $S_u$. Of course, this particular domain is unwieldy since it may consist of thousands of faces.

2. *Reparametrizing over a simple triangular base complex $K_\triangle$:*

Using the parametrization work of Eck et al. [2], Step 2 automatically constructs from the initial mesh $M_0$ a simple base complex $K_\triangle$ and a continuous map $\rho_\triangle : K_\triangle \to M_0$.

More precisely, the base complex $K_\triangle$ is computed using the connectivity of a *Delaunay-like partition* of $M_0$ (Figure 4) normally consisting of well shaped triangular regions. Additional care is thereby taken to achieve the topology of $M_0$ while building $K_\triangle$.

As the construction of $\rho_\triangle$ exploits the mathematical framework of harmonic maps, the parametrization $\rho_\triangle$ tends to have low metric distortion. The parametrization of $P$ from Step 1

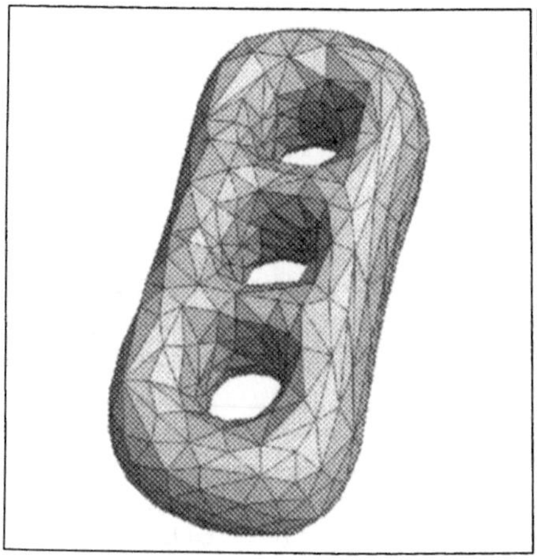

Fig. 3 Step 1: Reconstructed initial mesh $M_0$

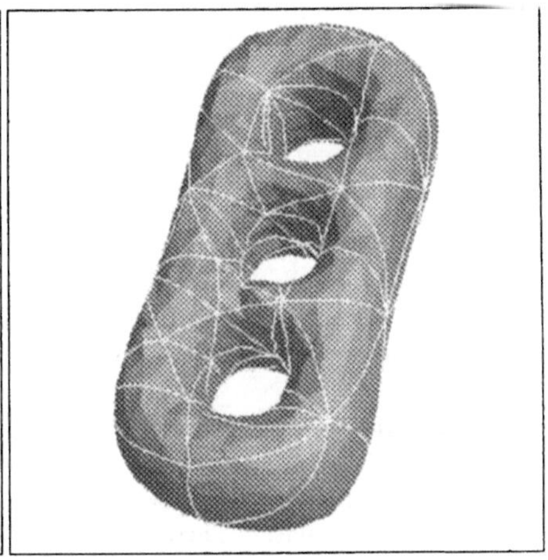

Fig. 4 Step 2a: Delaunay Triangulation of $M_0$

are mapped through $\rho_{\triangle}^{-1}$ to obtain new parametrizations over $K_{\triangle}$. This process is illustrated in Figure 5, where a line segment is drawn between each data point $\mathbf{p}_i$ and its parametric location on $K_{\triangle}$. Note that no geometric embedding of $K_{\triangle}$ into $\mathbf{R}^3$ is defined but has only been created in Figure 5 for illustration purposes.

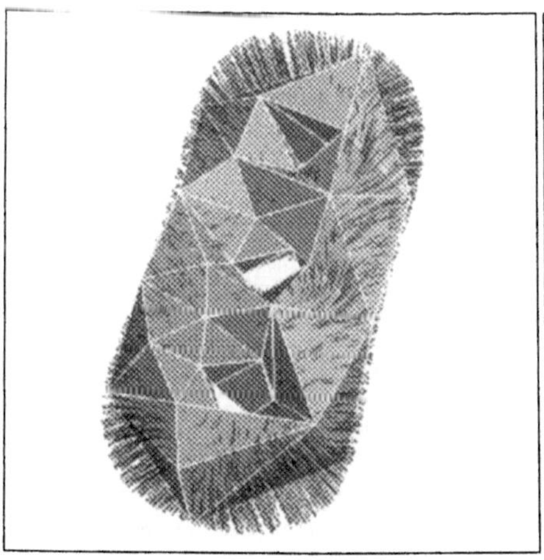

Fig. 5 Step 2b: Triangular base complex $K_{\triangle}$

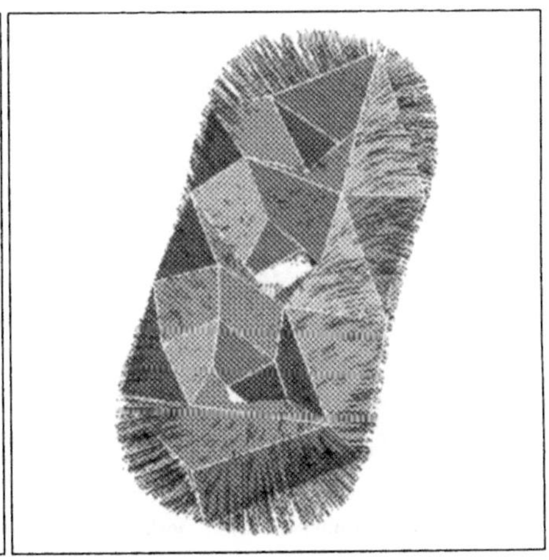

Fig. 6 Step 3: Quadrilateral base complex $K_{\square}$

3. *Reparametrizing over a quadrilateral domain complex* $K_{\square}$*:*

By merging faces of $K_{\triangle}$ pairwise, Step 3 constructs a new base complex $K_{\square}$ whose faces consist solely of quadrilaterals (Figure 6). The merging process is cast as a combinatorial graph optimization problem, whose goals are both to find a maximum pairing and to minimize parametric distortion. In case this process cannot find a complete matching, a global subdivision step would be necessary.

Finally, harmonic maps are again used to find a good reparametrization of the points $P$ from $K_\triangle$ to $K_\square$. The final parametrizations are again illustrated by the line segments in Figure 6 where we have created an embedding for $K_\square$ in $\mathbf{R}^3$ for illustration purposes only.

4. *B-spline fitting:*

Using the surface splines scheme of Peters [6], Step 4 constructs from the four-sided base complex $K_\square$ a control mesh $M_x$ which in turn defines a network of tensor product B-spline patches $\mathbf{s}_f$ that collectively form a $G^1$ B-spline surface $S$ of low polynomial degree ($2 \times 2$ resp. $3 \times 3$).

Fitting $S$ to the points $P$ is cast as an optimization problem over the vertices of the control mesh $M_x$, and is solved by alternating between a linear least squares fitting step and a parameter correction step. One problem with surface fitting is that the resulting surface may have unwanted "wiggles". Therefore the fitting term is augmented by an additional fairness term. Here, the ***thin-plate energy*** is used in a very similar setting as described by Dietz [1] in the case of fitting a single B-spline patch.

The result of the fitting step is shown in Figures 7 and 8 which was obtained in roughly 15 minutes on a 105 MHz HP 735 workstation. The surface $S$ consists of 49 B-spline patches whereby the maximal pointwise error is just 0.75% of the object diameter.

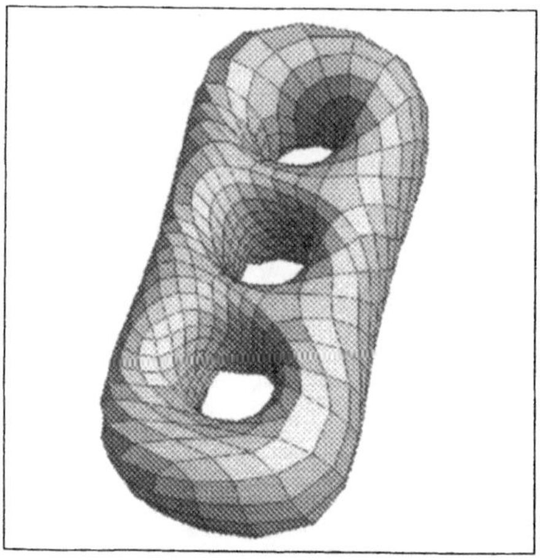

Fig. 7 Step 4a: Optimized control mesh $M_x$

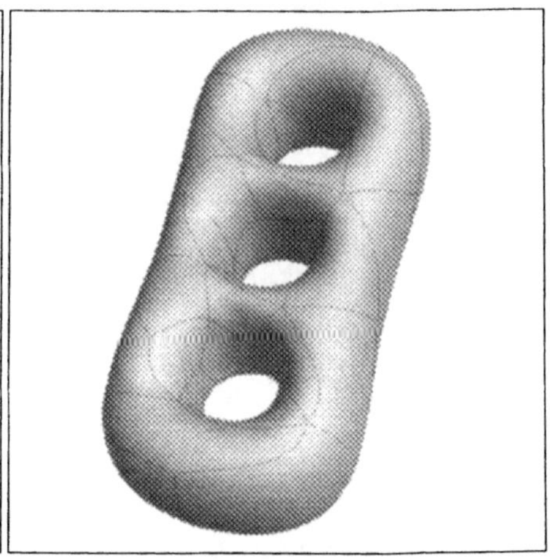

Fig. 8 Step 4b: B-spline surface $S$, 0.75% error

5. *Adaptive refinement:*

In order for $P$ and $S$ to differ by no more than a user-specified error tolerance $\epsilon$, Step 5 adaptively subdivides the faces of $K_\square$ into smaller quadrilateral surfaces based on the fit errors. After each step of subdivision, Step 4 is reinvoked to fit the refined surface. Further subdivisions are performed until the error tolerance $\epsilon$ is satisfied.

The result is a refined domain complex and a new control mesh $M_x$ (Figure 9) defining a new B-spline surface $S$ (Figure 10) within $\epsilon$ of $P$. Here, $\epsilon = 0.6\%$ implied 178 B-spline patches after 134 additional minutes of computing time.

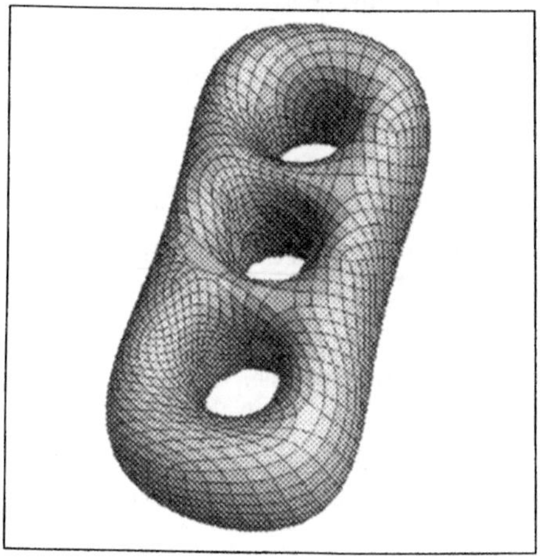

Fig. 9 Step 5a: Optimized control mesh $M_x$

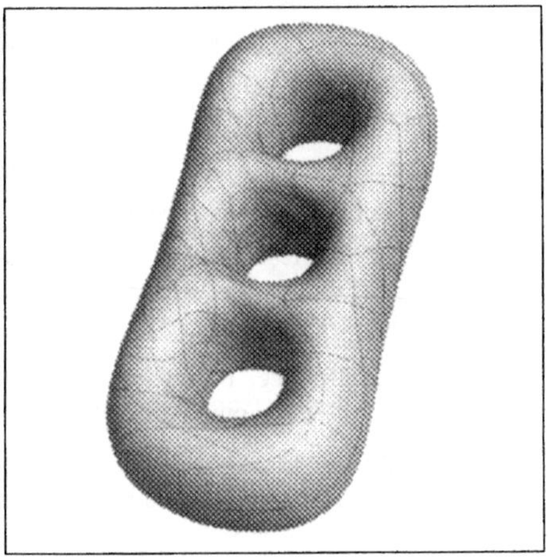

Fig. 10 Step 5b: B-spline surface $S$, 0.59% error

However, more impressive is the final example of 20,021 points of a human foot (courtesy of Technical Arts Co.). The surface in Figure 12, obtained in 33 minutes, represents obviously a reasonable fit but does not recover the fine details around the toes. Here, adaptive refinement with a prescribed tolerance of $\epsilon = 0.3\%$ gives a much better result after 228 additional minutes of computing time (Figure 13).

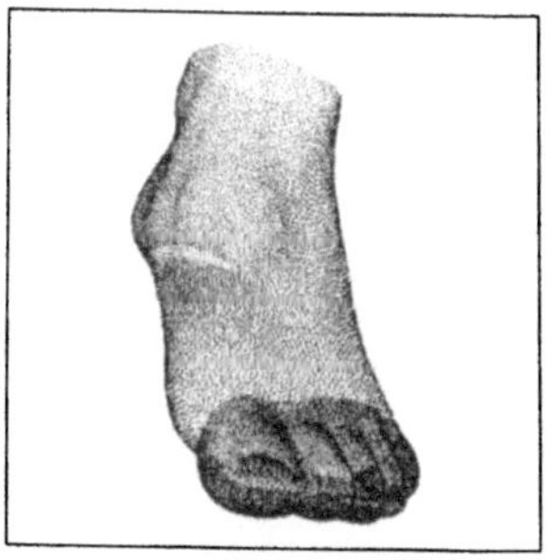

Fig. 11 $P$ (20,021 points)

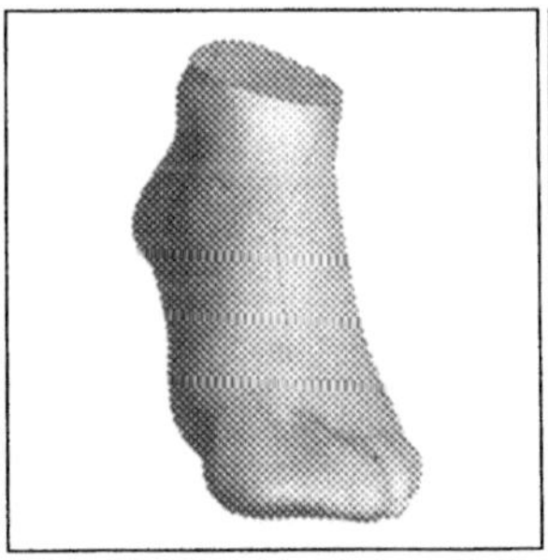

Fig. 12 29 patches, 1.2% error

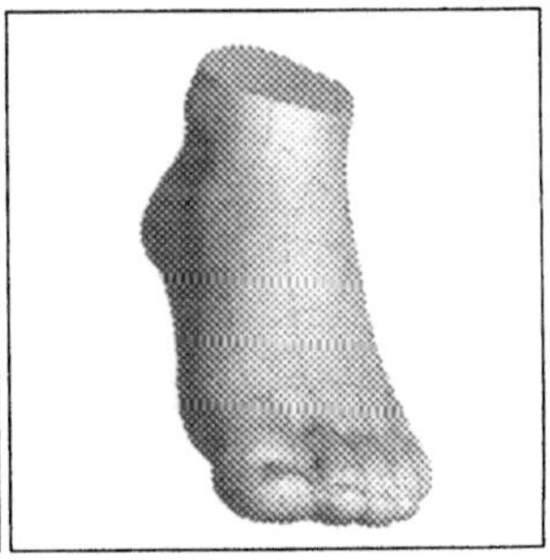

Fig. 13 156 patches, 0.27% error

# References


[1] Dietz, U.: *Erzeugung glatter Flächen aus Meßpunkten.* Technical Report 1717, Department of Mathematics, University of Darmstadt, Germany, February 1995.

[2] Eck, M., DeRose, T., Duchamp, T., Hoppe, H., Lounsbery, M. and Stuetzle, W.: *Multiresolution analysis on arbitrary meshes.* Computer Graphics (SIGGRAPH '95 Proceedings), pages 173–182, August 1995.

[3] Eck, M. and Hoppe, H.: *Automatic Reconstruction of B-Spline Surface of Arbitrary Topological Type.* Computer Graphics (SIGGRAPH '96 Proceedings), August 1996.

[4] Hoppe, H., DeRose, T., Duchamp, T., McDonald, J. and Stuetzle, W.: *Surface reconstruction from unorganized points.* Computer Graphics (SIGGRAPH '92 Proceedings), 26(2):71–78, 1992.

[5] Hoppe, H., DeRose, T., Duchamp, T., McDonald, J. and Stuetzle, W.: *Mesh optimization.* Computer Graphics (SIGGRAPH '93 Proceedings), pages 19–26, 1993.

[6] Peters, J.: *Constructing $C^1$ surfaces of arbitrary topology using biquadratic and bicubic splines.* In N. Sapidis, editor, *Designing Fair Curves and Surfaces*, pages 277–293. SIAM, 1994.


## Author


Dr. Matthias Eck
Küsterstr. 8
ICEM Systems GmbH
D–30519 Hannover
E-mail: me@icem.de

Dr. Hugues Hoppe
Microsoft
USA
E–mail: hhoppe@microsoft.com

# Two Methods for Surface Reconstruction from given Point Distributions

*Vibeke Skytt*

SINTEF Oslo

**Abstract:** Reverse engineering is the process of defining a CAD-model from an existing object. One of the stages in this process, is to define the surfaces within the CAD-model. This paper discusses two possible surface creation methods.

## 1 Introduction

In this paper two methods for surface reconstruction from given point distributions are presented, and a comparison between the methods are made. First a reverse engineering process is described, and in the next section we define the problem and the surfaces we want to find. Next, some requirements for surface construction methods are put up. The two surface construction methods are presented, a comparison is made, and finally, a conclusion is reached.

A reverse engineering process might be defined to start with a given point distribution. A set of data points are obtained, probably by measuring some object. Since the data points are measured, they are subject to errors. The magnitude of the errors depend on the measuring equipment.

We now assume that the data set consists of randomly scattered points, but we also assume the points to be explicit, that is the point set is projectable onto some plane.

Given a randomly scattered point distribution, we have to impose an ordering on the data set. This may be done by triangulation the points, for instance by a Delaney triangulation.

Next, the point set is divided into subsets where each subset of data points might be approximated by one single surface, and the surface boundaries are defined. Constructing good surface boundaries is a non-trivial task. Each surface should have a relatively simple shape, and it should contain no sharp ridges.

Once all surface boundaries are defined, we are able to build a topology structure, and to produce the surfaces. We now have obtained a surface model.

## 2 The Surface

The surface reconstruction from a simple subset of the original point set is the subject of this paper. We are now able to define our problem.

*Given a set of boundary curves and a sample of scattered data points lying interior to these curves, construct a surface that simultaneously*

- *blends (interpolates) the boundary curves*
- *approximates the scattered data points with an error less than a given tolerance*
- *is of good quality*

*Cross boundary information might also be given. In that case, the curves representing this information must be interpolated as well.*

We want to construct a tensor product B-spline surface as a solution to the problem stated above. Such a surface is described by

$$f(u,v) = \sum_{i=1}^{n}\sum_{j=1}^{m} c_{i,j} B_{i,k_1}(u) B_{j,k_2}(v) \tag{1}$$

where $\{c_{i,j}\}_{i,j=1}^{n,m}$ are the surface coefficients, $k_1$ and $k_2$ are the order of the surface in the two parameter directions, and $\{B_{i,k_1}\}_{i=1}^{n}$ and $\{B_{j,k_2}\}_{j=1}^{m}$ are the B-spline basis function defined over given knot vectors in the two parameter directions of the surface.

The spline space of which the surface belongs are defined by the polynomial order of the surface and the knot vectors.

# 3 Surface Reconstruction

We want to compare two different methods for surface reconstruction from given point sets. At the stage of surface construction, the original data set is divided into subsets, each well suited for being approximated by one single surface.

Before I start to describe the surface construction methods, I will put up some criterion on a good surface construction method:

**Flexibility** The demands on a surface are not constant, not even for different surfaces within one model. In an area with a lot of detail, accuracy is the most crucial aspect, while in other areas smoothness is more important. Thus, the user should have some handles to turn on to emphasize the most important criterion in each particular case.

**Stability** The surface construction method should produce sensible surfaces for all sensible point sets and preferably for some non-sensible ones.

**Good performance** We want an as good performance as possible, but this criterion still has got lower priority than stability, accuracy and smoothness.

**Accuracy** The method should have the ability to perform approximation within a given accuracy. Furthermore, the required accuracy should be satisfied with a minimum data size of the surface. The accuracy must be dependent on the measurement error. Requiring a more strict accuracy in the surface approximation than the exactness of the data points, is bound to give oscillating surfaces.

**Smoothness** We want smooth surfaces that are at least $C^1$ continuous. Sharp ridges in the data set are already taken care of by placing a surface boundary along the ridge.

I will return to these points later while evaluating the two methods, but first they need a presentation.

Both methods perform approximation of the data points. One of the methods, the two phase method, first generates data points on a grid and interpolates these points to produce a surface. The other method is more direct, it constructs a surface by solving a least squares problem with a smoothness term. The latter method is called a variational design method. The methods will be presented in more detail in the next two sections.

# 4 The Two Phase Method

In the two phase method, we assume that we have got a set of explicit scattered data points. The idea is to use these points to generate data on a regular grid. A tensor product B-spline surface is constructed by interpolating the grid points.

Generating data on a grid might be done in different ways. In our case we triangulate the data points corresponding to the surface, and evaluate the triangulation in the grid points.

The spline space of the surface is defined by those of the boundary curves. Interpolation of these curves is ensured by copying the coefficients of the curves onto the surface. Cross boundary information might be handled in the same manner.

Solving an interpolation problem over a grid leads to a set of smaller linear equation systems depending on the size of the grid. The grid size should depend on the data size of the boundary curves. All the equation systems are banded.

It is well known that interpolation might lead to oscillations and unwanted behavior in certain situations. To avoid this effect we do the following

- Avoid too much variation in the data set belonging to one surface. The point set should already be split in such a way that each subset is well fitted for being approximated by one surface.
- Evaluate a smooth function defined over the triangulation to produce the grid points, and not the triangulation itself. Evaluating the linear triangles of a Delaney triangulation directly may produce interpolation points that are not particularly well suited for its purpose. Instead we estimate a smooth function over the triangulation that interpolates the triangulation in its nodes, and evaluate this function.

In the two phase method, we use a special parametrization method where the idea is to obtain a parametrization of the scattered data points that is as close to chord length parametrization as possible. Perfect chord length parametrization is of course not possible for surfaces, but defining the subsets of points that the surfaces should adapt to, carefully, a good result is possible.

The first data set from the benchmarks of this workshop is used to illustrate the two phase method. See figure 1. As we can see, the final surface is defined on non-uniform knot vectors.

# 5 Variational Design of Surfaces

The variational design method for surface construction is a global approximation method. We want to create a smooth surface. A smooth surface should not have more bumps than what is necessary to satisfy the other requirements to the surface. It is likely that we want the bending energy of a surface to be minimized. Minimization of the bending energy could

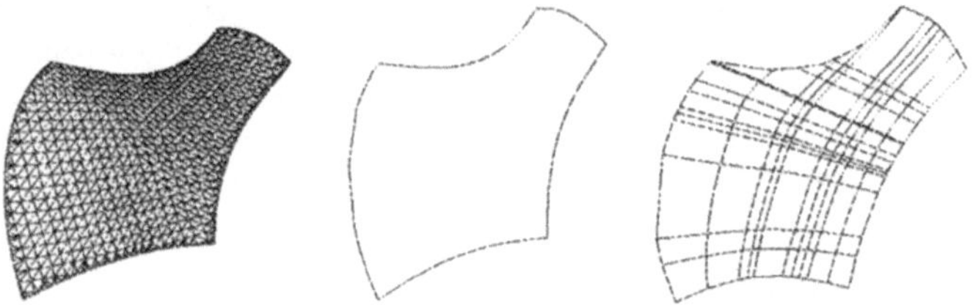

Fig. 1 First we see the triangulation of the data set, secondly the boundary curves, and at last the final surface.

be achieved by minimizing some measure for the curvature of the surface or some measure for the variation in curvature.

We want a smooth polynomial tensor-product B-spline surface, but spline surfaces are not particularly well suited for minimizing curvature etc. Minimizing these concepts leeds to non-linear sets of equations to define the spline surface, and solving these equation systems might be rather time consuming. Instead we choose to minimize another functional which is an approximation to minimization of curvature and variation in curvature. It also includes a term that minimizes area.

We want to find a spline surface defined by equation 1 over given knot vectors. $f$ will be found as a solution to the following minimization problem

$$\begin{aligned}\min_{\vec{c}}\Big[\int\int_{\Omega}\int_0^{\pi}\sum_{l=1}^{3}\omega_l\Big(\frac{\partial^l f(u_0+r\cos\phi, v_0+r\sin\phi)}{\partial r^l}\Big|_{r=0}\Big)^2 d\phi du_0 dv_0\\ +\omega_4\sum_{r=1}^{R}(f(u_r,v_r)-a_r)^2\Big] \qquad (2)\\ =\min_{\vec{c}}\Big[\int\int_{\Omega}\int_0^{\pi}h(u_0,v_0,\phi)d\phi du_0 dv_0+\omega_4\sum_{r=1}^{R}(f(u_r,v_r)-a_r)^2\Big]\end{aligned}$$

where $\{a_r\}_{r=1}^R$ denotes the scattered data points used to define the current surface and $\{u_r,v_r\}_{r=1}^R$ are the parametrization of these points in the parameter plane of the surface. The last term in equation 2 represent a least squares surface approximation scheme. The first part of this expression is a smoothness term which should ensure smooth surfaces also when there is too few or too badly distributed points for a pure least squares approach to succeed.

In our smoothness functional $\Omega = [a_1,a_2]\times[b_1,b_2]$ is the definition area of the surface $f(u,v)$. $\omega_1$, $\omega_2$ and $\omega_3$ are weights used to indicate the significance of the different terms in the smoothness functional while $\omega_4$ is the weight on the least squares approximation. At each point $(u_0,v_0)$ in $\Omega$, $h$ is integrated around the circle to ensure smoothness in all directions and not only in the parameter direction. See figure 2. Finally, $h$ is integrated over $\Omega$.

The three terms in the smoothness functional represents the following

- Minimize the first derivatives of the surface. This will imply minimizing the area of the surface. This term is included mostly for completeness.

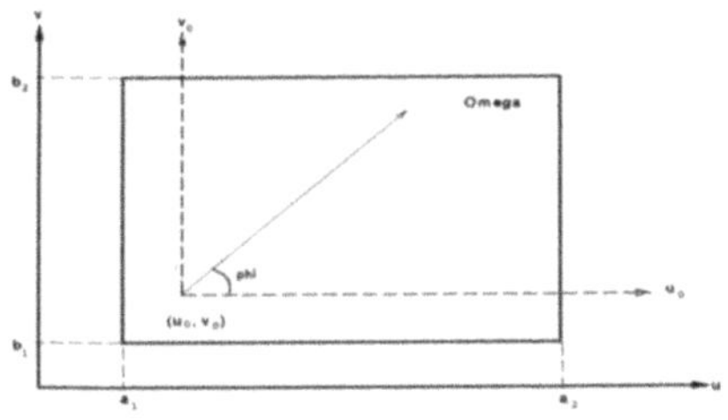


Fig. 2 Integrating $f$ at $(u_0, v_0)$ in all directions.

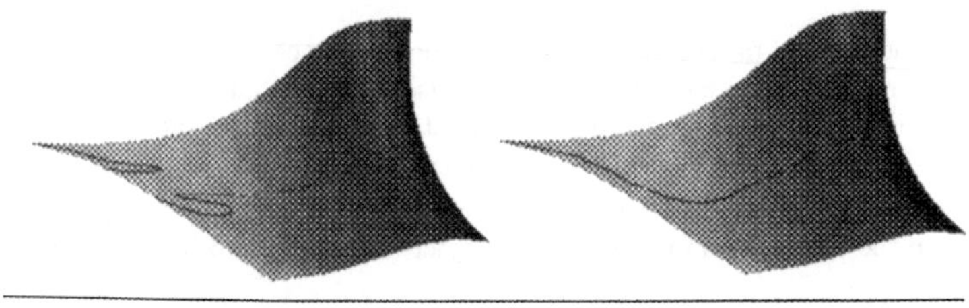

Fig. 3 Silhouette curves on the original surface and the modified surface.

- Minimize the second derivatives of the surface. This is an approximation to minimizing curvature. For reasons given in [7], it should not be the only term in the functional.
- Minimize the third derivative. This is an approximation to minimizing the variation in curvature.

These approximations to curvature and variation in curvature is good if the parametrization of the surface is close to chord length. Otherwise, the effects of the smoothness term is more unpredictable.

Figure 3 illustrates the effect of the smoothness functional. Silhouettes are computed on a surface made from scattered data. The silhouettes are found to be unsatisfactorily, and the smoothness functional is applied to the surface simultaneously with approximating the existing surface.

Differentiating the smoothness term, and elaborating the integral expressions, we end up with integrals of inner products of B-splines which may be computed exactly by Gauss quadrature. The minimization functional is transferred to a linear equation system where the number of unknowns is equal to the number of coefficients of the surface minus the number of coefficients that are fixed because of the boundary conditions.

The variational design method for surface generation is easily applied to non-projectable data points.

The exactness of the approximation depends on the data size of the surface, and on the weights. The weights give great flexibility in whether to emphasize smoothness or good approximation. These weights should depend on the magnitude of the error in the data points.

The method might also be used to change an existing surface if better approximation is

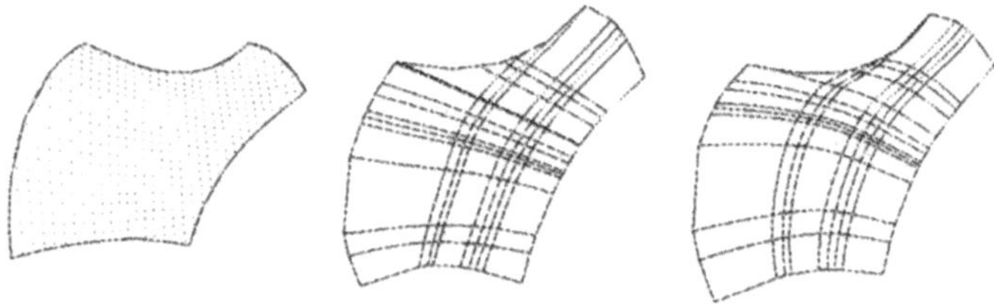

Fig. 4 A trimmed surface. First the point set and the boundary curves are shown. Then the resulting surface made by variational design and at last the surface made by the two-phase method, is shown.

required for one part of the surface, or if the data points in a certain area are of lesser quality than the other data points, and we want to decrease the importance of the approximation in this area. It is also possible to include approximation of another surface and approximation of the direction of the surface normal in a number of points.

Parametrization of the scattered data points is performed by making an intermediate blending surface from the boundary information and project the points onto this surface.

# 6 The Comparison

Our next step is to compare surfaces made by the two phase method and by the variational design method. Several benchmark examples were provided for this workshop, three of these data sets are suitable for being represented by one surface.

The first data set consists of fairly regularly distributed points with extra density along the boundaries. There are 5 boundaries so it is natural to produce a trimmed surface. The data set and the surfaces produced by the two methods are shown in figure 4. The surfaces look very similar, and the Gaussian curvature plots of the non-trimmed version of the surfaces, in figure 5 do not give a clear quality indicator either. Both surfaces change much in the curvature, but the size of the curvature is small as can be seen from the table below. In one area, the surface made by the two-phase method shows a better behavior. The accuracy is slightly better for the variational design surface.

The table below shows the number of coefficients for the surface in the 1. and in the 2. parameter direction. Then the maximum distance between the surface and the data set is displayed, and next the average distance. The minimum and maximum numbers of the Gaussian curvature are also shown. The number of points used to define the surface was 721.

| | No coef 1 | No coef 2 | max dist | average dist | min Gauss | max Gauss |
|---|---|---|---|---|---|---|
| VDM | 17 | 12 | 0.13 | 0.014 | -0.00115 | 0.00084 |
| Two-phase | 19 | 13 | 0.17 | 0.022 | -0.00138 | 0.00017 |

In figure 6 we have a look at the next data set. The data are clustered in some areas, and totally missing in others. Having data like this, it is a challenge to produce good boundary curves. The surface made by the two-phase method looks quite good, while the first version of the variational design surface has got bumps in areas where there is no data. There is too little weight on the smoothing term, and the least squares approximation to the data points,

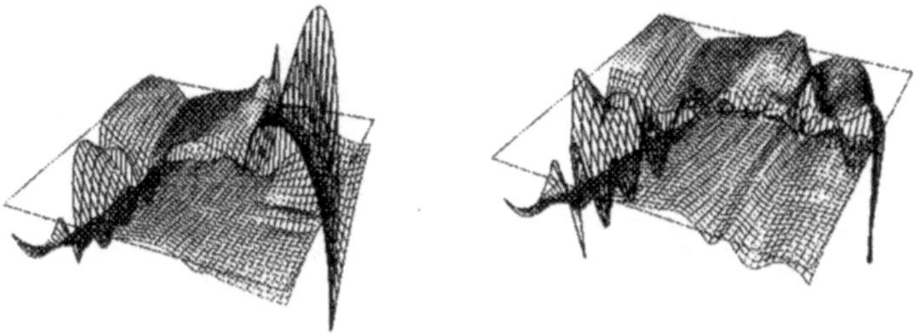

Fig. 5 Gaussian curvature of the non-trimmed version of surface 1. The curvature of the variational design surface to the left and of the two-phase surface to the right.

is unable to handle the surface creation on its own. In the next surface more emphasis is put on smoothing, and the results look better. However, the table below and figure 7 shows that there is a peak of curvature in one particular points at one of the boundaries of the surface. The accuracy is fairly similar for the two surface creation methods. In this case the two-phase method seems to be more stable, although it is clear that a lot may be done to the variational design surface by choosing the right weights and defining good boundaries.

The accuracy is not very good for any of the surfaces. A better solution for this point set is probably to divide it into several surfaces. The point at which the last variational design surface had its maximum curvature, was at a position well suited for division of the point set. The point set is, however, a nice exercise for testing surface creation methods.

The number of points used to define the surface was 1767.

| | No coef 1 | No coef 2 | max dist | average dist | min Gauss | max Gauss |
|---|---|---|---|---|---|---|
| Two-phase | 59 | 13 | 1.913 | 0.076 | -0.0179 | 0.0023 |
| VDM(1) | 58 | 10 | 1.885 | 0.082 | -0.3909 | 0.1105 |
| VDM(2) | 58 | 10 | 1.928 | 0.086 | -33.3 | 1.162 |

The last data set in this test consists of points belonging to one surface with a hole cut into it. Originally, it was about 100 000 points. The magnitude of the point set is reduced to 17485. The two surfaces look quite similar, and the table shows better accuracy for the variational design surface and less curvature for the two-phase method. The curvature plots cover the non-trimmed version of the surface.

The number of points used to define the surface was 17485.

| | No coef 1 | No coef 2 | max dist | average dist | min Gauss | max Gauss |
|---|---|---|---|---|---|---|
| VDM | 31 | 29 | 0.555 | 0.0048 | -0.0062 | 0.00067 |
| Two-phase | 31 | 29 | 0.747 | 0.0092 | -0.0015 | 0.000067 |

## 7 Conclusion

Neither the variational design method or the two phase method as they are presented here, supports satisfying a prescribed accuracy. The data size of the surfaces and the quality of the

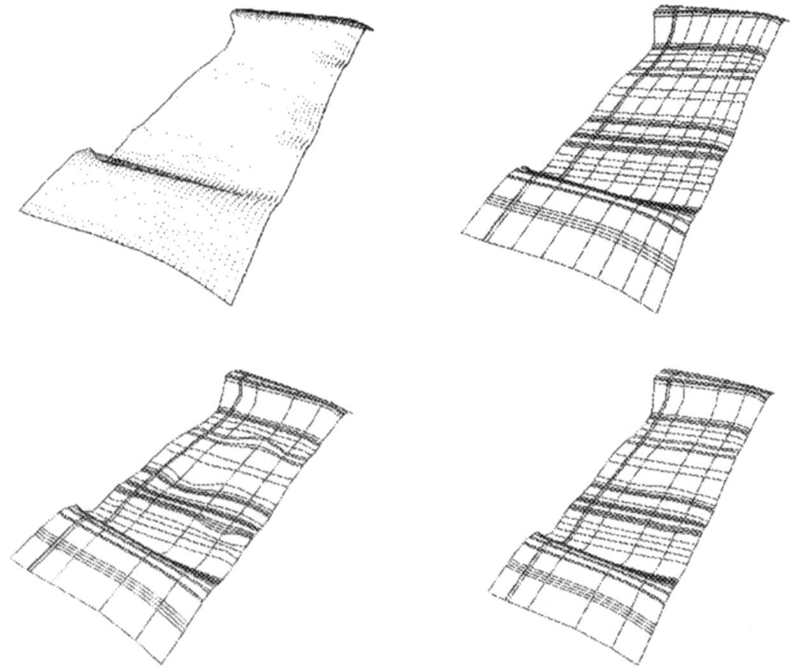

Fig. 6 A 4-sided surface. At the upper left, the point set and the boundary curves are shown. To the upper right we see the surface from the two-phase method. In the lower row we see two solutions using the variational design method. The surface to the right has got a higher weight on the smoothness term than the one to the left.

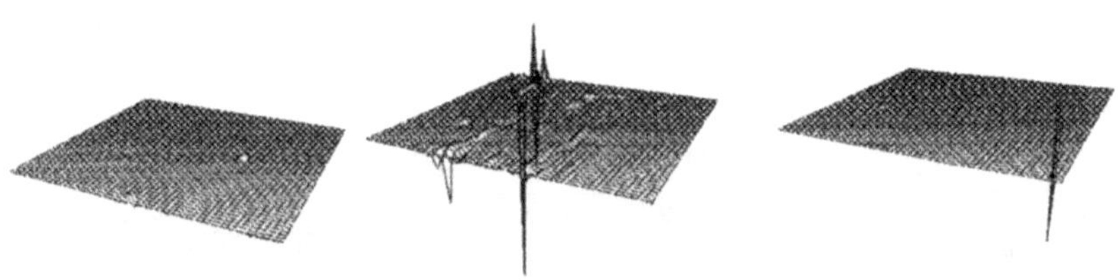

Fig. 7 Gaussian curvature of surface 2. To the left is the curvature of the two-phase method surface, then the two variational design surfaces. Note that the scale of the last surface is 1/10 of the scale of the others.

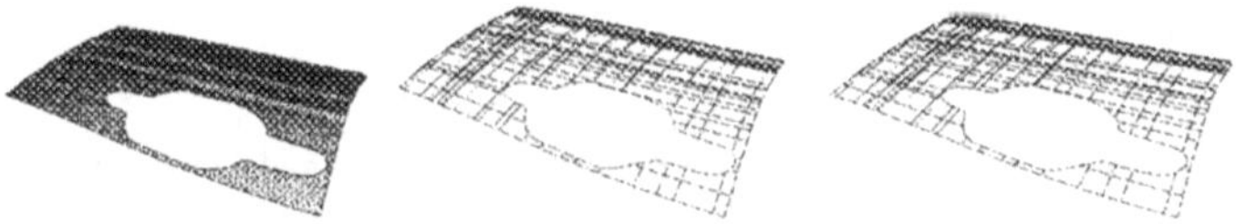

Fig. 8 A 4-sided surface with inner trimming. To the upper left, the point set and the boundary curves are shown. Next we see the surface from the variational design method and at last we see the solution using the two-phase method.

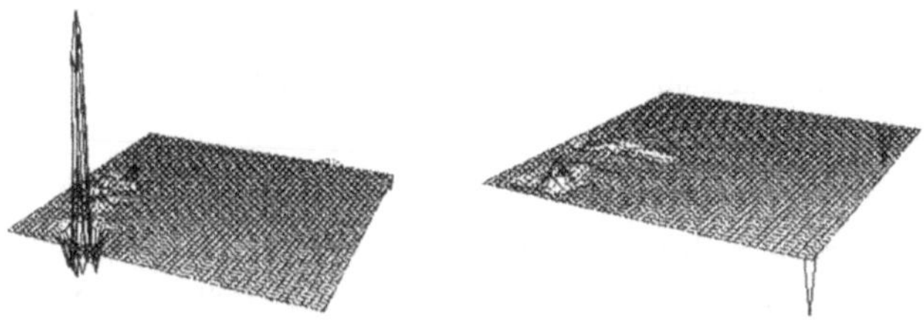

Fig. 9 Gaussian curvature of the surface from the variational design method and from the two-phase method.

boundary curves are limitations for the accuracy that is possible. For the variational design method, the weights influence greatly on the accuracy.

Returning to the criteria on quality for surface construction methods, we see the following:

**Flexibility.** The variational design method has most knobs to turn on and best possibilities for extensions. It is also a more direct method, and thus it is easier to see the consequences of changes in the input to the method.

**Stability.** The two phase method has less tendency to propagate a problem at the boundary into other parts of the surface. This is obvious when we look at how boundary information is included. For the two phase method, problems at the boundary will lead to rapid changes in the surface towards the boundary. For the variational design method, such problems might create oscillations in a larger area if boundary information is contradictory to the point information, and this is a global surface creation method.

**Performance.** The two phase method in general has got better performance. This is in particular the case if the data size of the surface is large. Solving several small equation systems are cheaper than solving one large. In addition it is more expensive to put up the equation system for the variational design method.

**Accuracy.** There is a tendency that the variational design method achieves a better accuracy with the same amount of data than the two-phase method. The accuracy of course depend on the weights.

**Smoothness.** The two phase method in general produces smooth surfaces. The variational design method produces smooth surfaces if the parametrization is good and the weights are set correctly.

It is no clear conclusion about which method is the best of the two presented here. Different cases ask for different solutions. In general, we might say that the two-phase method is more stable and more secure in difficult cases. On the other hand, it is easier to fine tune the variational design method if for instance the requirements on accuracy is very strict.

# References

[1] E. Arge, M. Dæhlen and A. Tveito. Approximation of Scattered Data using smooth Grid Functions. SINTEF report, (1994).

[2] Courant & Hilbert Methods of Mathematical Physics Volume 1. John Wiley & Sons. (1989).

[3] Günther Greiner and Hans-Peter Seidel. Curvature Continuous Blend Surfaces. Proceedings of Modeling in Computer Graphics. Springer-Verlag, (1993).

[4] Michael Kallay. Constrained Optimization in Surface Design. Proceedings of Modeling in Computer Graphics. Springer-Verlag, (1993).

[5] J. Kåsa and G. Westgaard. Analysis of Curvature Related Surface Properties, (1996).

[6] Weiyin Ma and J. P. Kruth. Parametrization of randomly measured points for least squares fitting of B-spline curves and surfaces. CAD. Volume 27, Number 9, (1995).

[7] Even Mehlum. Curve and surface fitting based on variational criteria for smoothness. Report for Central Institute for Industrial Research, (1969).

[8] Larry Schumaker. Spline functions. Basic Theory. John Wiley & Sons, Inc, (1981).

[9] V. Skytt. Surface Editing. Proceedings for the Dedicated Conferences on Mechatronics & Supercomputing Applications in the Tranportation Industries, ISATA, (1994).

# Author

Dr. Vibeke Skytt
Forskningsveien 1, Oslo
P.O. Box 124, Blindern
0314 OSLO
Norwegen

# Panel Discussion: Quo vadis Reverse Engineering?

*Summarized by G. Podehl*

Universität Kaiserslautern

**Participants:**

| | | | |
|---|---|---|---|
| W. Dankwort | University Kaiserslautern | Germany | (chairman) |
| P. Lecland | Dassault Systems | France | |
| A. Massabo | MATRA DATAVISION | France | |
| I. Raasch | BMW AG | Germany | |
| N. Sapidis | Technical University Athens | Greece | |
| T. Weißbarth | Intergraph | Germany | |

**Abstract:** There had been three different groups attending the workshop: Application industry, software vendors and research institutes. The participants of the panel discussion were chosen to be representatives of these groups. In the past the two university people worked in automotive companies for several years.

## 1 State-of-the-art

The discussion revealed the state-of-the-art of reverse engineering, which still is mainly surface reconstruction, or how Prof. Varady called it "reverse geometric modelling". Reverse engineering today is only working with geometric models instead of real engineering models. The term reverse engineering itself has a negative touch, as it sounds as if this technique is only used to run corrections on a part, which some designer did not manage to build correctly on the first attempt. A closer look at history showed that present reverse engineering was triggered mainly by two things: First, 3D design on a 2D screen is a difficult task, and therefore stylists prefer to work on clay and then regenerate geometry. If there were a virtual clay, then people would do 3D design on a 2D screen. Right now it is just impossible to create nice look- ing surfaces on 2D screens. Second, the improvements of laser measurement systems pro- ducing millions of 3D points by scanning physical clay models forced the engineers to find a way to make these huge amounts of points accessible for subsequent design steps. Hence, reverse engineering is used to translate the stylists' clay work into a computer accessible representation.

## 2 Major Problems

There are different kinds of model representation on different levels. Processing and translating the representations causes problems. The main problem turned out to be the work with points which do not carry any information beyond pure 3D positions. Points do not carry important additional geometric information like surface normals or curvatures. There is no possibility of storing surface interrogation results like shading or reflection lines. And - of course - points do not capture design history, functionality, aesthetic aspects or the stylists intentions which would dramatically ease the surface reconstruction. Up to now, all semantic information which exists about the objects is not used for reverse engineering. Even those data which can be captured today (like FEM data) are ignored in the process.

Other technical problems which were discussed are the numerical accuracy of the used algorithms, the handling of error propagation, the surface representations, insufficient user interfaces and system incompatibility. Especially, the question whether a parametrization of 3D models as surfaces (Bezier, NURBS, ...) is really needed or not, could not be answered. Future representations could use completely different surface parametrization or even work directly on point clouds with some new tools. Perhaps even a new mathematic approach must be developed to avoid the numerical problems of real-number computer algebra. It is also possible, that representations will be found, which do not need parametrization as they are used today (for example: voxel techniques).

## 3 Requirements and Future Trends

The future development in reverse engineering should be able to capture far more aspects of the products, than it does today. Instead of just focusing on geometry, reverse engineering could be used to support the entire industrial development an production process from conceptional design to manufacturing by capturing more high level semantics. The systems must be able to keep in mind how the user did construct certain things (an "associate-graph" ), because this is a very important piece of information and has nothing to do with the parts' rep- resentation. For the future the geometric models must be improved to create engineering models which have a history of creation and functionality; at the same time reverse geometric engineering must be extended to be able to store design intents and design process data. One way to achieve this could be the "feature based" approach with a global data structure (product model).

The trend in industry goes to avoiding physical models, but the present reverse engineering does not sufficiently meet the industrial requirements. The users do not want to work with mathematical terms like energy or different degrees of continuity. What they want are tools easy to use, individually customizable and powerful enough to manipulate the models globally and, even more important, locally. Stylists would like to use their tactile sense for judging sur- faces even when they are working on virtual CAD clay. So the user interface is very important for the success of reverse engineering tools. The system should be able to capture, what the user wants to do and offer him the possibility to individually define his own tools (user defined tools). Therefore reverse engineering research must be very pragmatic and understand itself as being a user driven discipline always asking the users, what they really want.

A system of the future should be able to automatically come up with some results after the user gave it some desired targets (i.e. functionalities and geometries). Of course he can not be sure about the quality of these results, but it will help the engineer to fulfil his targets. This means that the user would be able to control the design process by altering the targets and not by altering the intermediate results. The user should be able to go back in design history and alter all parameters the way he wants, because targets are not uniquely defined and the system must be able to satisfy different views from different users. A good way to achieve this would be the concept of building small, modular, interchangeable tools to cope with the changing requests of the users. Different algorithms from different sources and different ven- dors could then be incorporated into the model (managed with OLE, CORBA, etc.). The appli- cation companies would gain great flexibility from such developments.

## 4 Cooperation

One question which arose at the end of the discussion was, which role the universities could play in the reverse engineering area. Often research institutes suffer from the fact, that they do not really know, what the users want and what the vendors' problems are. Cooperation between software vendors and universities is difficult, as there is very much confidential information which cannot be communicated. So universities should focus on single hard problems which are too expensive to solve for the vendors or improve present algorithms. If an agreement is found of how to keep confidentiality, common projects could be started on problems with a high risk. The role of the universities should be the open minded trial of new ideas, where industry and software vendors too often only look at their daily problems.

## Author

Dipl.-Ing. Gerd Podehl
Universität Kaiserslautern
Lehrstuhl für Rechneranwendung
in der Konstruktion
D-67653 Kaiserslautern

# Benchmarks for Reverse Engineering

*Josef Hoschek*

Technische Hochschule Darmstadt

For checking the capabilities of the CAD–systems and the effectivity of the used approximation methods we have proposed some benchmarks. The goal was to approximate these set of points by one or more Bézier– or B-Spline surface patches with a distance error as less as possible (for instants 0.05mm) and a data volume as small as possible. The examples were developed by

- Daimler (Ulm),
- Starrfräsmaschinen (Rorschach)
- Holometric Technologies GmbH (Aalen).

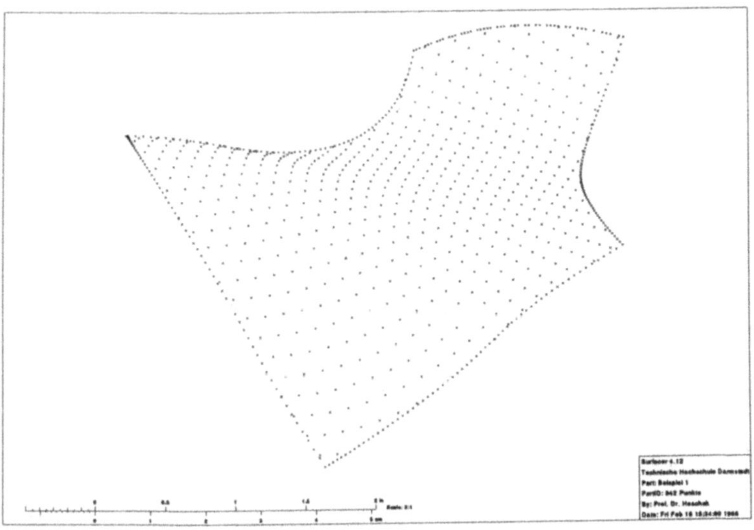

Example 1: Trimmed surfaces with 850 points on a regular grid and 5 digitized boundary curves.

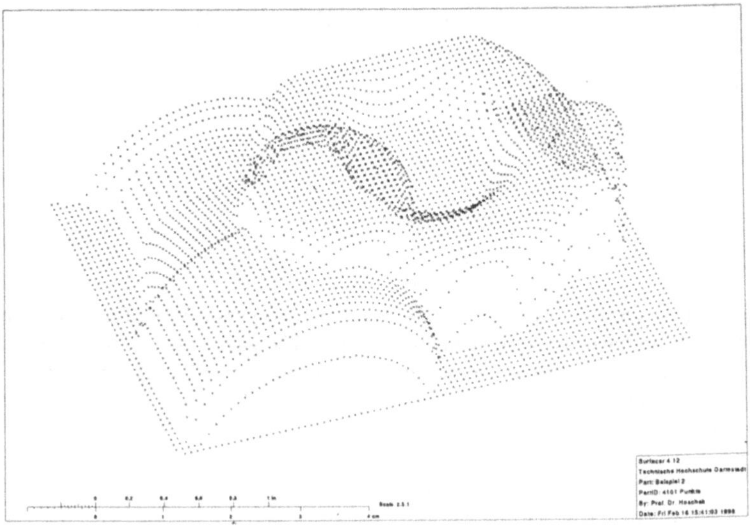


Example 2: Mechanical model with 4100 points on a regular grid. Because of the steep sides some parts of the model have low points densities.

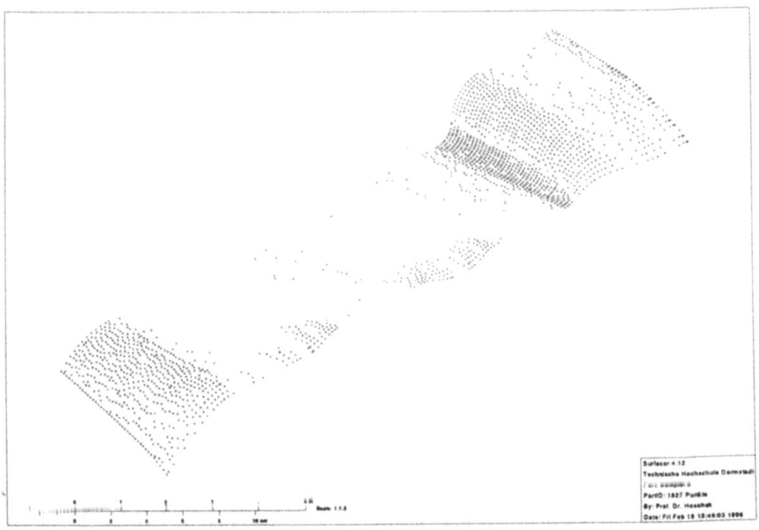


Example 3: metal sheet with 1827 points on a grid and some flat parts without measurement points.

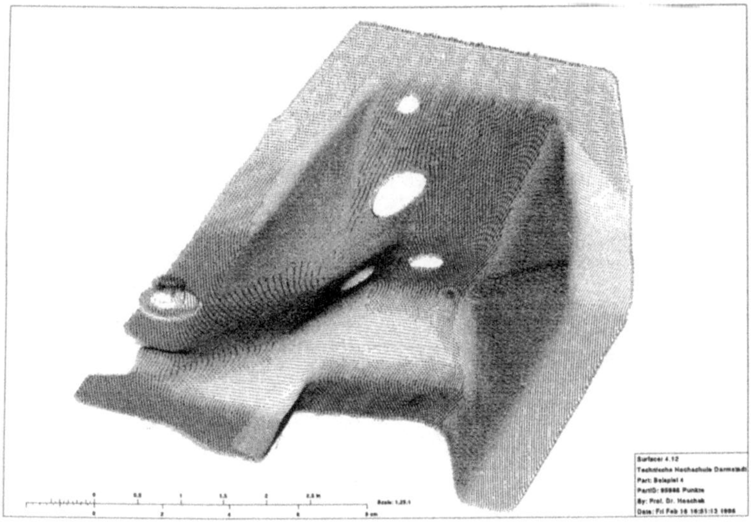

Example 4: Part of carbody with scattered 29097 points (from a laser scanner) and some drill–holes.

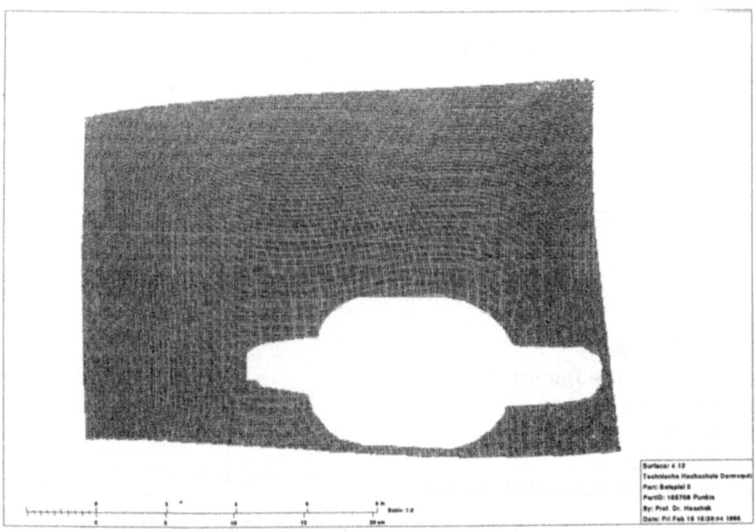

Example 5: Part of a car door with scattered 105700 points (from a laser scanner) with one hole.

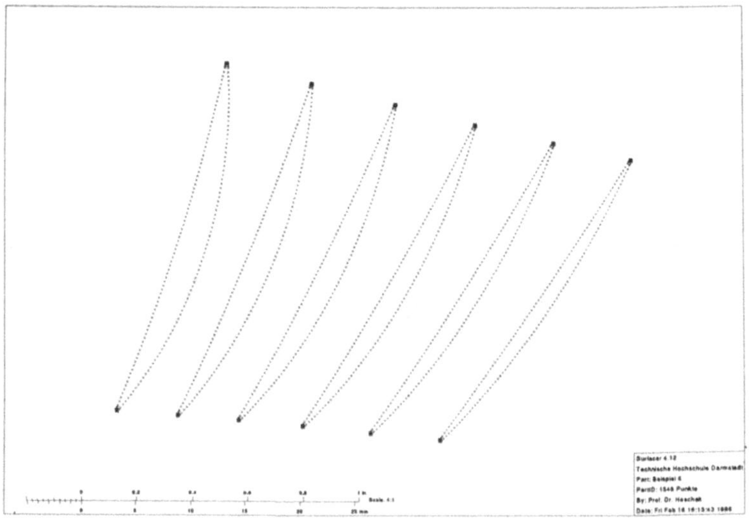

Example 6: Turbine plate with 1546 regular points on 6 profiles.

The benchmark data can be copied from the ftp–server of the Department of Mathematics at the Technische Hochschule Darmstadt with the following calling sequences:

```
ftp    ftp.mathematik.th--darmstadt.de
login:    ftp
password:    ftp
cd    pub
cd    CAGD
bin
get    bench.tar.Z
```

## Author

Prof. Dr. J. Hoschek
Technische Hochschule Darmstadt
Fachbereich Mathematik
Schloßgartenstr. 7
D–64289 Darmstadt
E–mail: hoschek@mathematik.th–darmstadt.de

GPSR Compliance

*The European Union's (EU) General Product Safety Regulation (GPSR) is a set of rules that requires consumer products to be safe and our obligations to ensure this.*

*If you have any concerns about our products, you can contact us on ProductSafety@springernature.com*

In case Publisher is established outside the EU, the EU authorized representative is:

Springer Nature Customer Service Center GmbH
Europaplatz 3
69115 Heidelberg, Germany

**Batch number: 10577936**

Printed by Printforce, the Netherlands